# Urban Systems Engineering: An Introduction To Urban Networks in Northeastern USA

## By Donald JG Chiarella, MSTM, CISM, CDMP

# Urban Systems Engineering

Cover art – A picture of one of the bridges in New York City. A network of bridges are located up and down the I-95 corridor on the northeast coast. Video cameras watch the bridges daily operation and state engineers respond to incidents at the bridges. Together, these transportation and electronic systems are urban systems engineering at it's finest.

Dedication – I would like to dedicate this book to my father who grew up in New York City and told me stories of his childhood in Brooklyn. He attended Alexander Hamilton High School. He joined the Air Force in 1952 and trained in radar at Kessler AFB, Biloxi Mississippi. He also earned distance education college credit from Cleveland Institute of Electronics. He retired in 1974 and worked 15 years for the Navy at St Indigoes, Webster Field as a GS-12 Project Engineer. I entered the USAF Academy to be like him. I studied Urban Planning and Information Systems Management at Maryland because of his careful guidance. I wanted to know how he grew up in the big city since I had always grown up on military base communities. I will always be indebted to him for showing me the way. He is my hero. I love you pop. Strength and courage.

ISBN 978-0-557-19367-7

Table of Contents

Table of Figures

1. Definition of Urban Systems Engineering

The definition of Urban Systems Engineering is a combination of the three words used in today's city environment. The size of cities has exploded and with this explosion has come new technologies that have drastically changed the way we live and think about city dwelling. The electronic systems advances have had great impact on all of society and the military. The city environment is no exclusion to this electronic systems explosion. The Brooklyn Polytechnical University and Steven Institute of Technology have new degree programs in Urban Systems Engineering. I looked at the curriculum and saw that it is primarily an urban technology program. At Maryland in the late 1970's my studies covered both areas or Urban Planning and IFSM. There was no Urban Systems Engineering major at that time. Both of these universities are located in the New York City area. I have been recently recruited by both of them. In the Northeast United States most innovations in electronics are marketed in the New York Area for large sums of revenue from the millions of people living there. With successful marketing in these areas comes the expansion of electronics networks and devices in the northeastern United States. Public agencies also have contributed significantly to the new electronic networks in this region.

A "network" can also be a transportation network or other type of network for the purposes of this book. There exist "networks" of "networks" which complicate the problem. Simply stating that a network runs through Manhatten may not be descriptive enough. Many networks run through Manhatten. A bus network, a taxi network, a telephone network, a sewage network, a telecommunications network all run through Manhatten at the same time. Some of the networks converge and diverge. Some of the networks are intertwined in the same conduits and trenches under the city. The transportation network is intertwined with the video network of cameras watching the traffic patterns in downtown NYC. The city government has staff watching the traffic patterns in this manner. It has long been known here in Maryland near Baltimore that if you want to do a big networking project look at how New York does it and copy them. The 5 Burroughs are large enough that no other city on the east coast will be able to handle more capacity than New York City. This means that whatever technology you develop in your state or city you can be sure if it has been running in New York City that many of the bugs will be fixed and problems will be ironed out before you try to implement solutions in your state or city. The size of the problem is just huge and complex for New York City. Maryland politicians have also copied systems from New York City (Compstat – Citistat - Statestat) for crime prevention and other public safety and funding tracking functions. Never underestimate the abilities of people to borrow   ideas from good working sources. Even with that many people do not like dwelling in the city (of any size). The city poses unique challenges to us (crime and poverty) and our systems and networks help rationalize the association of elements we are managing in this world.

So "Systems Engineering" is not a brand new term anymore. It means engineering of systems of hardware and software to control other systems. Almost similar to computer engineering except that there are many types of systems. A naval vessel may be a system that has systems engineers working on it. An aircraft may have systems engineers working on it. A city has systems working in it and thus we have "Urban Systems Engineering". An amalgamation of systems and networks of systems in the urban environment. What defines urban? The census bureau tells us that we track data in what is called the SMSA or Standard Metropolitan Statistical Area. Therefore the New York City SMSA includes all the people in the 15-30 miles radius from downtown Manhatten. New Jersey would certainly be in the SMSA. It may extend all the way to Connecticut in the north. This is merely an example of the influence of a major metropolitan area on surrounding areas. Other cities may have larger or smaller sizes of geographical measures of the SMSA. It is enough to realize that the definition has changed over the years and cities have become larger and larger – onto megalopolises[1]. Soon there will be the megalopolis city BosRich – Boston to Richmond. As development fills in the suburban and rural areas between the cities this is the megalopolis affect. And as development fills in there will be more and more Urban Systems required for sewage, telecom, telephones, highways, neighborhoods, storm drainage, parking lots, buses, railways, and traffic lights. There will be one continuous city from Boston to Washington mainly accessible by I-95 and the other local and state highways traversing it. A huge number smaller communities and town are located in the BosWash megalopolis and there are even regions and airports for what used to be the smaller SMSAs. Yes

---

[1] Frenchman Jean Gottman, Megalopolis, 1961 first introduced this concept of a huge city from Boston to Washington. The Greeks originally named a city "Megalopolis" but it was very small in geographic size.

this is where we are going. I have watched it for over 30 years since college. Travel along the I-95 corridor tells you that this indeed the case. The only real difference is that the states have kept their tax areas constant thus defying the realities of the urban development the past 50 years. States even have anti-sprawl and anti-leap frog development plans such as Smart Growth in Maryland. But it is too late to turn back the clock. Urban Systems Engineering is what we see in the megalopolis of BosWash in the form of flight patterns, Amtrack, I-95, and HDTV and cables lines. We are an interconnected Northeast USA and it is always growing bigger. Managing that growth is daunting.

Some states have adopted the Harvard Smart Growth Policy of managing growth and sprawl in suburban areas. Leapfrog development is not allowed and strict zoning laws prevent bad land use. New Towns[2] have been allowed to flourish in formerly rural areas. The new towns are built around malls and residential neighborhoods. Transportation routes redefined their boundaries and farms became residences of home developments. They are also called "Planned Communities" and "Satellite Cities". I would liken them to a military base community in that they are self contained and do not depend on other cities. They also have all the amenities of nice communities. This all comes at an expensive price tag and there is a problem with affordable housing in these "New Towns". There may be an extra tax called a "Community Association fee" which feeds the organization required to manage day to day issues like trash collection, snow removal, home design changes, etc. I know because I served on the

---

[2] Greenbelt (1940's) and Columbia (1968), Maryland and Reston (1970), Virginia.

Architectural committee in Columbia's King's Contrivance Spring Breeze neighborhood from 1991-1994.

A lot of growth in the New York metropolitan area can be attributed to one man – Robert Moses. Moses was the most powerful man in New York from the early 1930's through the 1960's. He directed many urban projects including highways, the UN building, bridges, and tunnels. He was the architect of modernizing New York to accommodate vehicle traffic. His methods were not always friendly to the people of New York but his accomplishments were many. He was so good that other cities like Baltimore, Portland, Chicago had him give them advice on city planning. He even ran for governor of New York in 1934. He was a juggernaut of urban planning in the 20th century.

2. The expanding SMSA and Micro SMSA

The 2010 Census is almost upon us. It will yield much larger sized SMSA's than ever before. It is almost a certainty to show that the Northeast is becoming one major Urban area in the country. The blackout of 2003 proved that this electronics grid is vulnerable to catastrophic effects of complex networks of electronic generation and software errors in the system. There is a new definition of micro Standard Metropolitan Statistical Areas outside of major metropolises. This has occurred on the Eastern Shore of Maryland now called a micro SMSA by the 2000 census results. But population does not tell the whole story, it merely measures the current capacity and density of an area of people that we may call city or urban area. The SMSA is the way we know we have large urban areas in our region. Two SMSA's may intersect like Washington and Baltimore. Together they create the Baltimore-Washington region or suburban Maryland.

The size of the SMSA may grow between census decades or decline. In general, Baltimore has seen more people leaving the city for the suburbs thus the population of the city has decreased but the size of the Baltimore SMSA is increasing due to rise of the suburbs population. Various counties populations are increasing over time. There are 4 big suburban counties in Maryland that hold the overflow of people from the cities. These are Prince George's, Anne Arundel, Baltimore, and Howard Counties. Life in these counties is within the SMSA of either city. Anne Arundel also has the city of Annapolis.

Planning in the SMSA of Baltimore is done by the Baltimore Metropolitan Council (BMC) and planning in the Washington SMSA is done by the Washington Council of Governments (COG). They provide the regional transportation planning for the areas. Other SMSAs in America have similar transportation planning groups. The key is that all these government planning groups have to talk to one another to phase in any new projects in the region that cross political boundaries. The local, state, federal, and county governments all play a part in the regional planning process of the SMSA. Often a major transportation or telecom project will be discussed in these council meetings to ensure everyone is informed about the status of the projects. Funding of the projects can come from many sources also. Federal funds can be used for Metrorail, interstate highways, airports, seaports, and some IT projects. All of these are Urban Systems. States usually try to match a percentage of funds on certain projects. Grants also help pay for major projects.

Figure 1. Northeast area map

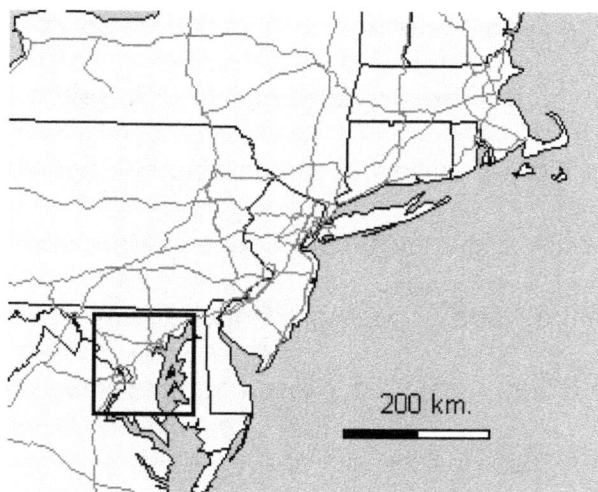

200 km.

Baltimore- Washington SMSA
**1990**

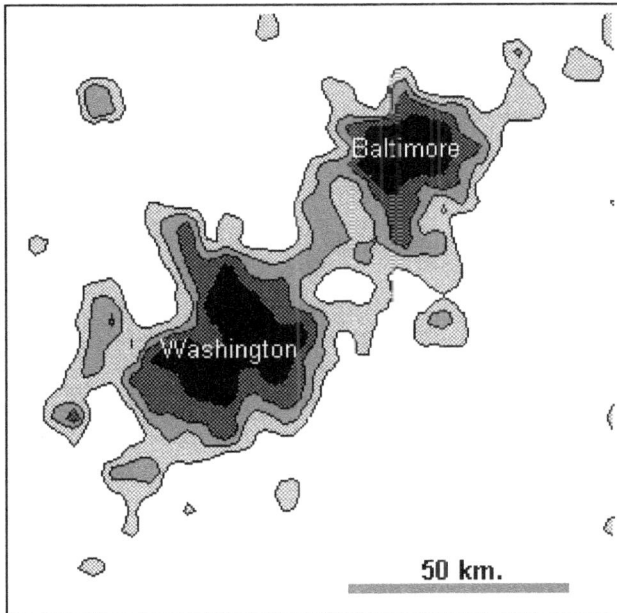

Figure 2. Baltimore-Washington Density Map

3. Environmental Considerations

Many of the states in the Northeast are water bound states by the Atlantic or Great Lakes. Many of these states have mountain ranges and seaports in the same state. This unique combination of topography creates a climate that simulates most climates in the United States. The environment coexists with the urban areas and expansion of urban areas requires environmental impact statements on major projects affecting wildlife and natural habitats. Civil engineering projects such as highways require environmental approvals before launching.

States such as Pennsylvania and West Virginia have mining operations in rural mountainous areas that have special regulations that restrict urban development. States such as Virginia, Maryland, Delaware, Massachusetts, New York, and Maine have extensive fishing and seagoing trade. Any water runoff into the tidal waters must be clean and free from bacteria. Otherwise the aquatic life forms in these waterways would die. Engineering projects often leave toxic waste and garbage behind to really foul the environment. Man made problems need to be diminished before we create our own early Judgement Day[3]. We need to be optimists and act with new programs to save the environment.

Urban systems engineering affects all the environment. As evidence take the Chesapeake Bay watershed area. Any building projects in the SMSA of

---

[3] See the last chapter of this book for more on this.

Baltimore and Washington will cause certain toxins to head downstream into the Chesapeake. Homeowners in the SMSA are restricted from using too much fertilizer on their yards because runoff ends up in the bay. Some five states empty streams and rivers into the Chesapeake. The Susquehanna, Potomac, Shenandoah, and Delaware rivers all empty into the Chesapeake Bay. We must be cautious of environmental laws in building the urban systems infrastructure. The Hudson river watershed has a similar problem passing from Lake Champlain through Vermont, upstate New York and New York City. Never has man so needed to keep his water and air clean for future generations. This is a public responsibility and people need to use common sense in handling of toxic waste materials. We have to ensure that raw sewerage is processed properly and not dumped into the rivers. Factories must clean up operations that may dump toxins into the rivers in the watershed areas. This can only be regulated by government control. Industry has proven time and again that they can not regulate themselves[4] when it comes to polluting the environment.

---

[4] The Democratic Party is more regulatory based in environmental policy than the Republicans.

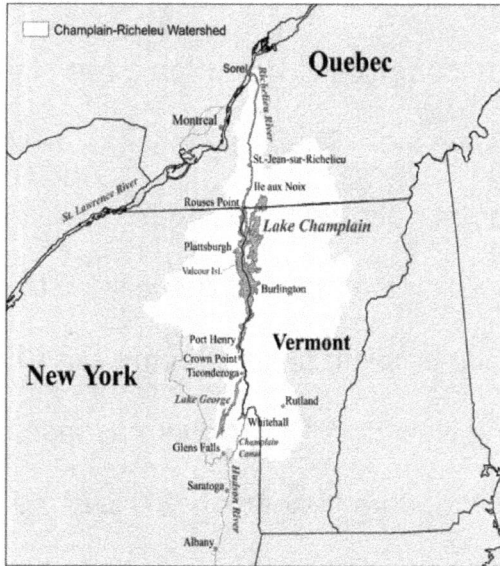

Figure 3. Lake Champlain Watershed    Figure 4. Chesapeake Watershed

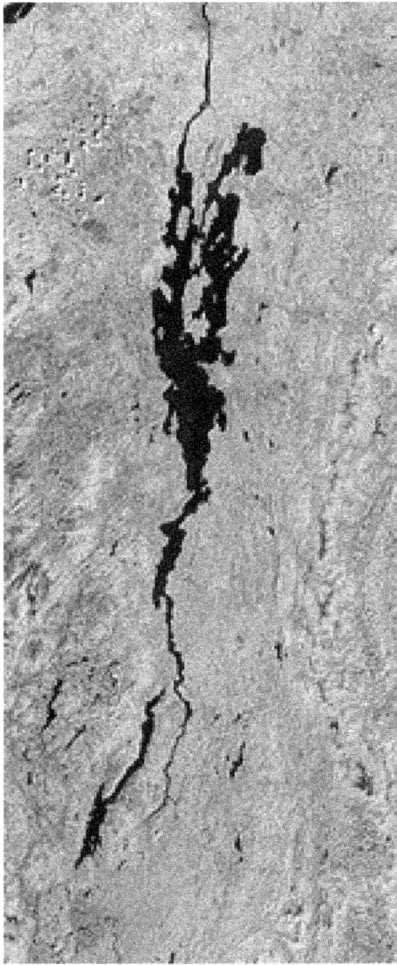

Figure 5. Lake Champlain          Figure 6. Chesapeake Bay Topographical Map

So as new green technologies arise we have the opportunity to make sure our environments are cleaner for future generations. Clean energy alternatives will create a better environment. Nuclear power plants will provide clean power. Nuclear waste needs to be dealt with at the national level. The problem there is that no state wants to be a dumping ground. Yucca Mountain[5] in Nevada was

---

[5] Congress can not decide on where to put nuclear waste if not at Yucca Mountain. For now some nuclear sites store their own waste. It takes 10,000 years for radioactive materials to cool. This problem is paramount to creating electricity with nuclear power.

the place previously designated to store nuclear waste for 10,000 years but that

is now endangered in Congress.

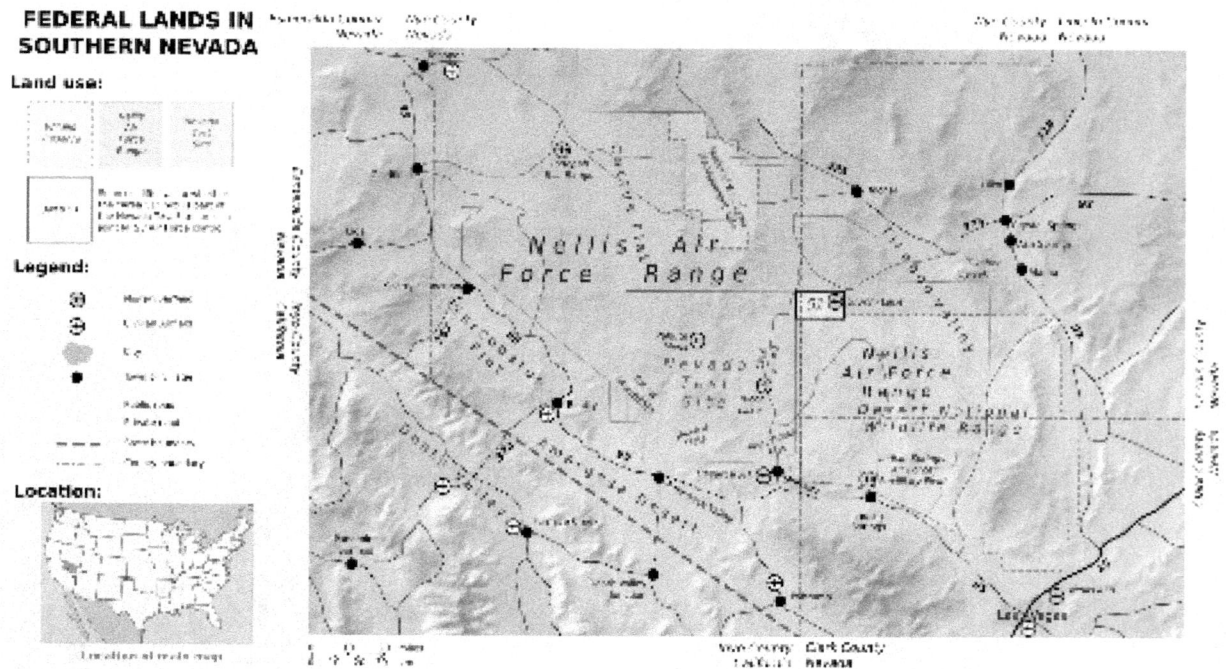

Figure 7. Land Around Yucca Mountain, Nevada

## 4. Electrical Systems

The primary infrastructure engineering achievement of the 21st century is the US power grid[6]. No nation is the world can boast a system that generates the power that the United States generates on a daily basis. We are a modern society that runs by electricity. Consider if you could live like the Amish do without any modern conveniences such as electricity and running water. I think most of us would agree this would be a very difficult task.

Local housing codes help distribute the power grid to the consumer. Costs of power consumption are skyrocketing in all parts of the country. Seniors must have Air Conditioning in the summer in many parts of the south. In the Northeast, we have particularly harsh winters that require more power. The blackout of 2003[7] proved that the power system is not infallible in the Northeast. An error in Ohio propagated through the system and blacked out New York City and other areas of the northeast.

---

[6] Indeed electricity was discovered and invented by Benjamin Franklin a American Scientist. He gave his findings to the French in exchange for the French Fleet support at Yorktown.

[7] The primary cause was a branch fallen on c wire that shorted and propagated through the northeast systems shutting them down one by one. It was not terrorism as some stated at the time. The age of the system is in question today.

Figure 8.2003 Northeast Blackout[8]

## Major cities affected

| City | Number of people there |
|---|---|
| New York City and surrounding areas | 14,300,000 |
| Greater Toronto Area (Golden Horseshoe) | 8,100,000 |
| Newark, New Jersey and surrounding counties and suburbs | 6,980,000 |
| Detroit and Surrounding Areas | 5,400,000 |
| Cleveland and Greater Cleveland | 2,900,000 |
| Ottawa | 780,000 of 1,120,000* |
| Buffalo and Surrounding Areas | 1,100,000 |
| Rochester | 1,050,000 |
| Baltimore and Surrounding Counties | 710,000 |
| London, ON and Surrounding Areas | 475,000 |

---

[8] Accessed from Wikipedia website.

| Toledo | 310,000 |
|---|---|
| Windsor | 208,000 |
| Estimated Total[20] | 55,000,000 |
| | |

The power grid controls urban systems engineering projects. Without power the systems are useless. Good old 220 and 120 volt AC power. DC power is good too. Hopefully we will see the promised revolution in solar and fuel cells battery power. Nuclear power contributes the most to power generation. Much power is still generated using coal. The whole push towards clean energy will be one which helps us clean the environment while adding to power resources.

What types of urban systems rely on electrical power? Almost all of them. Traffic lights, video cameras, computers, networks, telecom lines, and many others depend on electrical current. In the last few years we have seen many traffic devices that are solar battery operated independent of the electrical grid. This is a good thing for electrical systems that can stand alone. As long as there is sun power there will be regeneration using solar power. I think the sun has an estimated billion more years of life so we should be ok there.

The electricity grid has been named one of the targets of terrorists and is subject to protection from cyber attacks on control computers called SCADA[9]. If the terrorists can knock out these control computers then they can knock out the power that we rely on in the modern world. This is a big part of the critical infrastructure protection effort by the FBI and DHS.

---

[9] Supervisory Control and Data Acquisition computers that run the nuclear power plants.

## 5. Telecommunications

The micro computer revolutionized the way we communicate. Telecommunications companies are large entities providing a valuable service to customers and households. They compete for services and yet they know they will not go out of business. Telecom hotels house T1 lines in large cities. The ISP[10] you choose will determine how well your computer connects to the outside world. If you pick a dial up service you will pay less but have slower service at 56K baud rates. If you choose a DSL line you will be faster but also pay more. If you choose broadband fber you will have a very fast connection. If you choose satellite connection you have the fastest connection. All of these are available today in many areas of the Northeast. Typical costs range from $10 to $150 a month depending on your service type. There are bundles of service that provide for telephone, internet, and HDTV connection all on the same input line. These are very good and give you a discount for all three services. Again there is also the Magic Jack you can buy from TV commercials for VOIP[11] free telephone service. The companies like Comcast, Verizon, Cox , and Dish Network play an important part in building the network infrastructure. Without their investments the networks would not be as well connected or as large as they are. They own the networks.

---

[10] Internet Service Providers such as Verizon, Netzero, People PC, and others.

[11] Voice Over Internet Protocol.

The old AT &T telephone network is just the basic elements of the huge telecom network in the Northeast. Interestingly enough, when President Lincoln installed the telegraph it was between Baltimore and Washington so the telecom history goes back to the civil war in this region of the country. The government regulates telecommunications in the FCC or Federal Communications Commission. The code of federal regulations or CFR is the book that contains the regulations. The NTIA has produced a chart called the Frequency Allocation chart which shows the radio frequencies in the United States. GSA also has an important role in managing government computers and telecom contracts and the process of contracting. These agencies are located in Washington DC.

New York has the stock exchanges which require massive computer power and networks. There are miles of wire in the stock exchange to support computer operations. The reliance on networks is phenomenal. Traders in the whole country network into the system on a daily basis. Traders from overseas in other world markets also network into the system. This network is a financial network and has sometimes put the whole exchange in jeopardy. Computer trading can be dangerous if the computers automatically sell off at one time. The basic idea is buy low sell high. But when everyone is taking profit there is the risk of a crash. This was demonstrated in the movie "Trading Places". I think that today anyone can be an investor and many people track their investments using computer networks. This is an easy way to gain intelligence into the problem of investing or guessing where the market will go in the next few months or years. Probabilities play a part also in the future growth of stock, mutual funds, and other assets. The combination of computer power and investing tools is an idea that has come of time. Professional investors definitely are using their computers to provide a service to the public. In the Northeast the trader's network is full of people in the investment industry.

The SEC or Securities and Exchange Commission is the federal agency that regulates investing and commerce. They use a system called Edgar to monitor investing and daily transactions at all the American Stock Exchanges. They are accountants and they are very good at finding white collar criminals. They often work with the FBI White Collar Crime Division.

## 6. The Smart Building

Architects are building better buildings all the time. Their designs are becoming more like artwork than traditional Greek buildings. Each new building has the latest in modern electronics and is considered a smart building if it provides for telecommunications networks wiring inside the walls. Most new buildings are wired this way with Ethernet or some other network. Now wireless networks may also be employed in smart buildings. Smart buildings are also able to lock down in case of fire and extinguish the fire. They allow human to interact with the systems placed in the buildings. Alarm systems are embedded in the smart building. Lighting is controlled by computers as is climate control. A group of smart buildings can work together to manage lighting in a shared garage. Street lighting may be controlled by the smart building. Parking lot lighting and security video cameras are also controlled by the smart building. These systems are standard in many of the new structures now built in the downtown areas of the United States.

A recent article in Computerworld stated

"As building automation systems (BAS) that control heat, air conditioning, lighting and other building systems get smarter, they're converging with traditional IT infrastructures. Emerging standards are enabling data sharing between building systems as well as with other business applications, improving efficiency and real-time control over building operating costs. Information security concerns, immature standards, the reluctance of vendors to give up proprietary technologies and ignorance among IT professionals of the convergence trend are all slowing the pace of this transformation, but it's gathering momentum.

Facilities managers are driving the change by demanding more-open systems. They're pushing BAS vendors to transform today's closed technologies into Web-enabled applications running over industry-standard IP networks. And the management of BAS is likely to increasingly fall to IT."

Jim Sinopoli wrote a book on Smart Building technology. In the Venn diagram[12] below he outlines the intersection of green and smart buildings. Note that the "green" component is energy efficiency and the "smart " component is electronic systems.

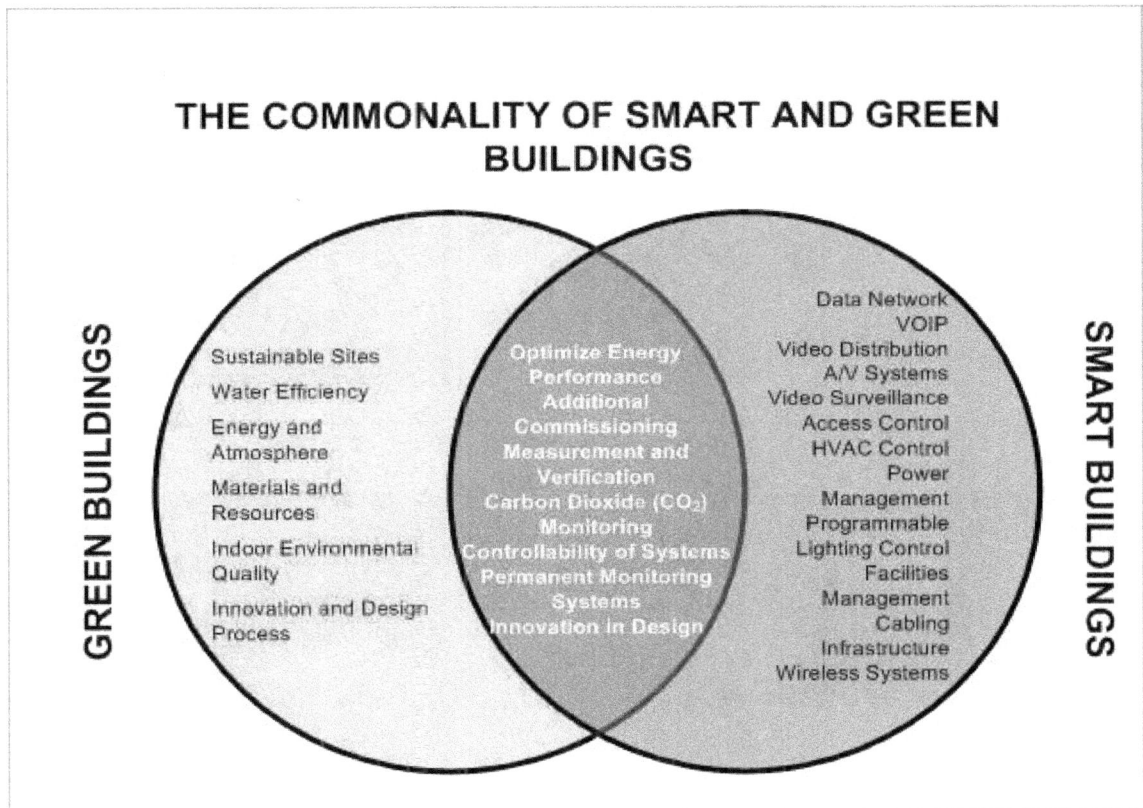

## THE COMMONALITY OF SMART AND GREEN BUILDINGS

**GREEN BUILDINGS**

Sustainable Sites

Water Efficiency

Energy and Atmosphere

Materials and Resources

Indoor Environmental Quality

Innovation and Design Process

Optimize Energy Performance
Additional Commissioning
Measurement and Verification
Carbon Dioxide ($CO_2$) Monitoring
Controllability of Systems
Permanent Monitoring Systems
Innovation in Design

Data Network
VOIP
Video Distribution
A/V Systems
Video Surveillance
Access Control
HVAC Control
Power Management
Programmable Lighting Control
Facilities Management
Cabling Infrastructure
Wireless Systems

**SMART BUILDINGS**

---

[12] Sinopoli, Jim, "How can a green building also be a smart building?" in Dec 2007 "Smart Buildings Magazine accessed at www.smartbuildings.com on Sept 11, 2009. Jim also wrote a book on the subject.

Figure 9. Venn Diagram of Smart & Green Buildings

## 7. Solar Power Sources

Independent solar power has always offered an alternative to the power grid as we know it. Private home owners in some parts of the country have invested in solar power to augment their power intake. We have not instituted solar power collection in government agencies yet. Government buildings are ideal for solar power because they are managed by the same entity. The movement towards solar power must start with government tax credits and incentives to people to buy the technology.

Figure 10. Solar Collage[13]

---

[13] Accessed from www.BuilditSolar.com.

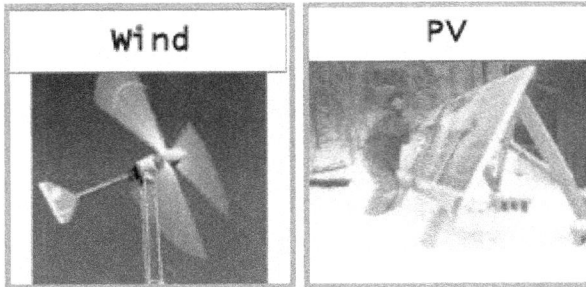

Wind    PV

     I think many people do not trust solar power yet which explains why they have not invested in solar systems. You do not have to buy a complete solar power system. You can buy a smaller system to heat your water only. You can buy passive solar windows in your home that bring in maximum solar energy to reduce heating.  You can ensure your home is built facing the south so you gain solar energy through the rear windows all year round. You can plant leafy trees to the south for shade in the summer and sunlight in the winter.  Dome homes and underground homes offer even more insulation in winter. Then you can mix them with active and passive solar power. These are all design considerations of the new home builder.

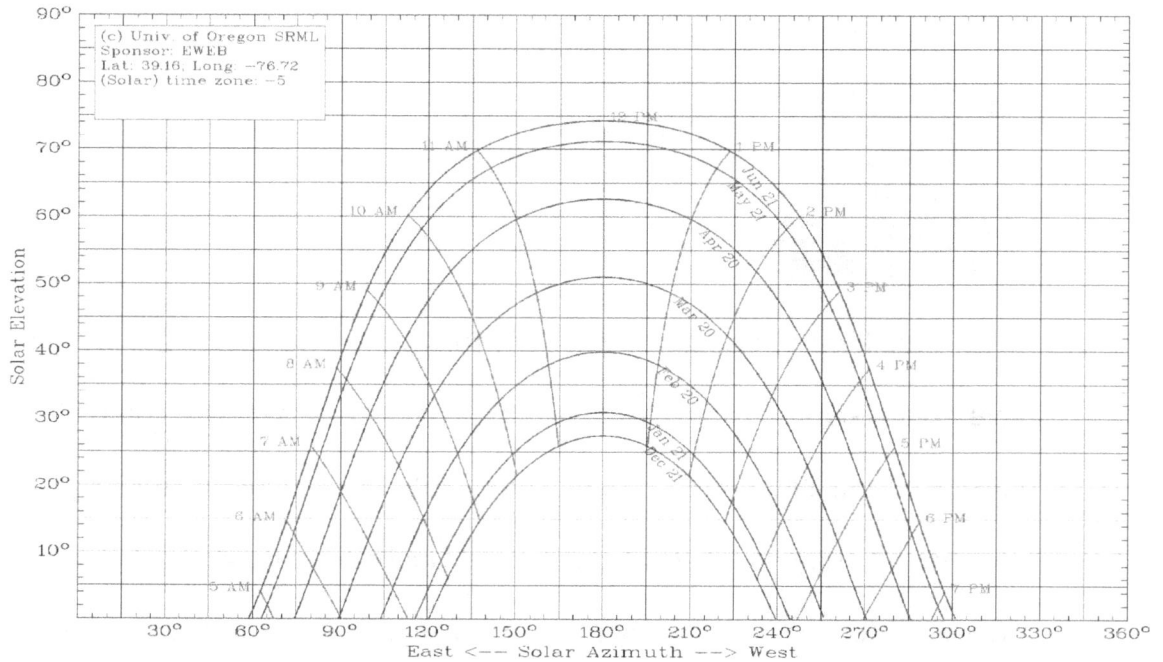

Figure 11.  Solar Chart for Zip Code 21076 – Hanover,  Maryland - BWI  Airport[14]

The solar chart above indicates the percentages of sunlight during the various times of the year at the location specified.  This chart is for Hanover, Maryland next to BWI airport.  Notice that the sun angle for December 21 is only 28 degrees compared to 75 degrees for June 21.  The farther south you go in the United States the higher the angle of the sun will be.  This means more powerful radiation through the atmosphere to the solar collectors.

If you live in a state such as Colorado, Arizona, or other southwestern states you get more sun days than other states. This is ideal for daily solar energy collection.  You can put more energy back into the grid if you run your

---

[14] Ibid.

own solar generator. If more people did this we would not have the high

dependence on electric companies that we do today. Look at

www.builditsolar.com for new ideas on how to build your own solar arrays at

home. The possibilities are tremendous for the homeowner. As we move

towards a national policy of clean energy the time was never better to invest in

solar energy for our homes. It may be that the politicians are waiting for the

industrious homeowners to get on the solar bandwagon and do the projects to

make them self sufficient.


I have only witnessed one solar building project in government in all my

years. This was at the Wyoming rest areas on Interstate 25. The entire rest area

building was a passive solar building. It had no heat or air conditioning and was

designed for sunlight heat. The other projects are solar run highway electronics

on the side of the roadways. These devices are often small and are self

contained. They are more widely used than ever before. The payback on solar

energy is immediate and there is also cost savings on energy generation.


At any rate the do it yourself kits for home solar systems are great but I

would suggest hiring a certified electrician to help you. He will understand how to

hook up the connections to your house power system better than you can. There

is an issue with voltage differences in solar systems and your home electric

circuit. An inverter may be required. You will have to have a permit and

inspection in most places. If there is one thing you do not want it is to ruin your power system in your house. Too much do it yourself can be bad. Trust me on this one I have ruined an entire basement and deck patio because I tried to go it alone.

## 8. Personal Devices

It started with the personal computer and has not ended. Electronics companies are banking on the personal device market. They make it seem like you can not live without a personal device and then they introduce a new personal device every few years. They forget to tell you about hidden costs of networks. The network companies even buy the hardware now a days for single fee. Cell phones alone are a multi billion dollar industry. I-pods, Blackberry's, Music machines and other personal devices are all designed to steal your money. Portable radios were quit the hit many years ago.

Now you have to have devices for everything. Why hasn't one company come up with a device that does everything? The computer industry gets a boost every time applications are written on cell phones. The line between personal computer and cell phone is very blurred. What they are calling cell phone applications are only a shell of what mainframe applications used to do. They are single function small applications. They are not multi functional applications. The general population has not caught onto this yet. And you must be careful about where you can get your cell phone reception these days.

There is not 100% coverage as the commercials would suggest. Rural areas may not have cell towers based on usage counts. You would be remise to count on cell reception everywhere in the United States even in remote areas. The urban areas will be fine but this is why I am writing this section. You need to know the limitations of personal devices and why they are making millions of dollars for their inventors – not you the end user. Annual costs and fees can be extremely high for cell phones. Many people are now stopping their landline services and just paying for cell phone service at home. This avoids double charging and ensures coverage in the house by at least one phone system.

## 9. Vehicle Devices

New vehicles are loaded with extras to increase the price of the vehicle to the dealer. Some of these electronic devices are nice but rarely are they all required. The newer GPS locator system[15] can navigate you to your destination without much trouble. They can even route you so that you pay minimal tolls on a toll highway. The vehicles have radar to detect when you are close to another car and some can even apply brakes. Some vehicles can determine if you are veering off course or out of your lane and send you an alarm for corrective action. The smart highway tested in California can control cars speeds on the highway in a platoon that runs at high speed. The highway has sensors in the roadway that signal the car. Cars can even communicate with each other to avoid crashes. Police radar detectors tell us when we can miss a ticket. Radar jammers can interfere with the police radar signal. Some states have outlawed these devices. Automobiles are just smarter than they used to be thanks to the computer revolution. As the expense comes down we will see more computer controlled car systems in the future on normal vehicle models. Today the luxury models have all the devices and the highest prices.

The DOD has run an experiment with the radio controlled vehicle and they now have ones that have passed the Baja desert test track. Unmanned vehicles provide a great service to the military. The software runs the car with sensors

---

[15] Tcm Tom is an example of this type of GPS in-car system.

and video cameras for remote operation. This is very similar to the unmanned aerial vehicles (UAV) now in production for the Air Force. Imagine the uses of an unmanned vehicle to the civilian population. Ambulances, police cars, fire trucks can all report to duty without a driver. The possibilities are endless for the robotic car or truck. Cars will be able to go in places where they previously could not with a driver. The remote driver will be able to view the vehicle path from a remote site like the UAV pilot. How exciting a time for new car owners?

One author predicts the robot cars will replace mass transit. They will also lower accident rates and accident costs. Robot cars will obsolete the need for oil by running on electric power. They will refuel and park on their own. This is a huge change from the way we operate vehicles today. This is similar to the robotic vehicles network in the futuristic movie "Minority Report".

Cars will come with new types of power systems and energy usage. The compressed air car was developed several years ago in France. It ran entirely on cans of 200 pounds of compressed air. The hybrid technologies will give way to new technologies. Electric cars are on the highway. The Smart car company is a type of small electric car. The problem is that it is not crash worthy. The smaller the car the greater the risk of injury in accidents. This is the rub with new electric cars all of which seem very, very small. With the government takeover of Detroit car makers there will be more and more new ideas on the roadways in the future. Japanese car companies will also develop new ideas. They are very competitive with gas engine models and can do well in the hybrid market as well.

DARPA held a competition called the Urban Challenge in 2005 and 2007. Perrone Robotics entered Tommy Sr. and Tommy Jr. autonomous vehicles in the competition. They were able to complete the challenge with minimum errors. Tommy Jr. cost $60,000 to build from off the shelf parts and computer components and laser sensors[16]. The system that runs the car is Sun Systems Solaris OS and Java located on a computer board in the vehicle. Lasers look out from the vehicle and tell how close it is to other vehicles and geometrics on the roadway like curbs and center lines. There is no driver and no controlling human being. The mission path was programmed into the vehicle 5 minutes before the competition. The prize money was 2 million dollars offered by DARPA. This means robotic cars are now viable. The DOD needs these vehicles in situations where troops will not have to go into harms way. Tommy Jr. is a Scion with racks of sensors on the exterior and computers in the trunk. Perrone also makes a UAV and BotKart robotic machines. This is a new arrival on the transportation scene.

---

[16] Accessed from www.perronerobotics.com on 1 Oct 2009.

10. Infrastructure Builds and Funding

Today the private sector owns the telecommunications infrastructure and the government owns the highway infrastructure. Citizens own the home and community infrastructures. There have been attempts to privatize the whole infrastructure. There have been attempts to governmentize the whole infrastructure. What is surely to happen is that there will be intertwining of the three infrastructures. Complex systems require complex infrastructures and this can very expensive. Infrastructure funding and protection can also be expensive. There could be a day when the FCC and the other regulatory agencies such as DOT need to come together for airlines networks and vehicle interstates infrastructure building. There is no one way to look at it anymore. Computers changed all of that. Advanced systems engineering is required of all these systems for them to work in unison.

Funding for projects will have to evolve to allow cross funding of different infrastructures and projects under different titles. Strict adherence to regulations may not allow the best systems solutions when systems are integrated with more than one component. For example, airlines systems need to have FCC and FAA compliance.

Funds may come from private sector or public sector for highways. The adopt a highway is one example of a privately funded maintenance program for a section of highway. Private airlines foot the bill for commercial aviation but the federal government pays for military aircraft and operations. The railways are run by AMTRAK which is a public company. Other rail companies include Santa Fe, CSX, Burlington Northern, Union Pacific, Norfolk Southern, etc. The commercial boating industry is run by private companies too. State governments basically regulate transportation issues so there is state money and federal money in all modes or transportation. The money can come directly from congress and be matched by state funds or grants can pay for some of the costs. Grants do not have to be repaid. Federal USDOT appropriations also do not have to be repaid.

The IT industry relies on IT companies to pay to invest in IT. Agencies also invests in IT in government buildings. The internet was built and funded by DARPA in the early days but now most ISPs are private telecom firms. They are fee based firms making money directly from customers for services rendered using their equipment. That is, the telecom backbone is privately owned in America. This has been true ever since AT&T owned the first telephone lines in America after the 1934 Telecommunications Act gave the monopoly the power to wire America. The Telecommunications Deregulation Act of 1984 basically opened up the competition in telecommunications lines in America but the backbone was already in place by AT&T. Small companies could offer competitive prices against Ma Bell. The Baby Bells took their place in regional telecommunications. Today, internet runs mostly over new fiber lines or older copper wire. The fiber upgrades have cost companies the installation fees they charge to customers. Some internet can run over satellite also. Commercial HDTV runs on satellite. This is a direct payback scheme. The companies market directly to customers through TV commercials. The bundle package is common where a user can get telephone, internet, and HDTV service for one price. Some of the companies are Comcast, Verizon, and Cox. Vonage is also a competitor in the Voice over IP market. For $39 a person can buy Magic Jack and get internet VOIP telephone service for free. This appears to be a real deal since PC World magazine gave them an award.

11. Hoyt's Concentric Circles impact on Systems

Hoyt's concentric circle theory[17] discusses the rings around the central business district of the inner city. The theory states that development in each ring goes from urban to rural as you leave the CBD outward. The suburbs are the places people live in the circle around the city CBD[18]. Exurbia is the far suburbs which are far more expensive. What does this theory mean to systems engineering in the metropolitan area? It means that systems customers will be either businesses in the CBD or residents in the suburbs and usually both.

Take Verizon for example, the majority of their business customers may be located in the CBD with a few in malls and shopping centers in the suburbs. The majority of their residential services will be in the suburbs. Hoyt's model can predict which type of systems will be in what sectors of the city metropolis. Furthermore in Exurbia, systems customers may receive similar services to rural customers. Linking rural customers to urban customers can be a challenge for the company. There will be less payoff for rural customers but a full statewide service may be sought by the state government. This is the case in Maryland. Network Maryland has been a system of fiber optic cable that connects the Eastern Shore of Maryland (mostly rural) to the Baltimore-Washington metro areas network. The cost efficiencies by economies of scale in the urban areas

---

[17] From Urban Geography textbook from University of Maryland, 1978.

[18] Urban Geography, University of Maryland, 1978

are shared with the costs of wiring the rural areas.  New home developments in the rural areas come with fiber optic cable access as a package deal thanks to the cooperative efforts of the cable company in connecting rural and metro areas. Surely, if the company could they would focus their efforts on the higher returns of the metro areas.

Transportation networks like the metro rail lines may have originated in the CBD many years ago. But who would ride a metro system that does not go to the suburbs anymore?  More and more metro systems are being extended to beyond the suburbs to exurbia or way outside of the city beltway network.  This is true of both Baltimore and Washington rail networks. They even connect with the MARC train.   Hoyt explains the concentric circle sectors the trains pass through as they journey from CBD to CBD.

Airports are generally located in the suburbs because of the need for space. At one time LaGuardia airport was in the suburbs of NYC. But NYC grew larger and larger and new suburbs were created. Now LaGuardia appears to be downtown.  The same is true of Regan National Airport.  At one time it was located on the outskirts of Washington but now is considered downtown compared to BWI and Dulles International airports who also service the same area.   The FAA has a network of control towers all along the northeast.  The flights between areas are inter and intra regional.  Communications networks must be in place to support flight operations betweens these network nodes.

Hoyt lets us compare where we locate the airport compared to the CBD and waterways (for emergency crash landings).

12. HDTV - A Major League Solution

The HDTV revolution has taken urban areas by storm. We in the middle Atlantic have almost always received Turner network and the Atlanta Braves on our cable TV. Now we receive the Chicago Cubs and Chicago White Sox network WGN[19] as well as our own MASN sports cable channels. This year the MLB network started up and it has been a hit with baseball fans. They can go inside any ballpark in America on any network. What a wonderful change from viewing baseball in the past. It is as though we have been liberated from the single channel home teams viewpoints and can now cheer for the entire league. HDTV has expanded in many other ways in terms of the number of channels offered and the medium of choice. Fiber optic is now preferred to coax. The FIOS fiber network has more than 1500 channels. This is just amazing. Future investments in fiber will be worth the high quality pictures received. The current price is high but so is the quality. Fiber offers security and high signal strength. There are fewer corruptions of the signal than either satellite signals or digital coax. signals. There is no dependence on the weather like satellite signal reception. Someday the price will be down and more people will enjoy the large number of channels on FIOS. Comcast and Cox are also networks in our immediate region. They compete with Verizon prices but are not as technically advanced. The cable TV market in the Northeast is highly competitive. There are more than 50 million viewers here.

---

[19] WGN – Chicago, MASN – Middle Atlantic Sports Network

The HDTV requires an HDMI cable attached to the HDTV set. This costs a few dollars but it means the death of analog TV. A large screen systems can cost as much as $4000. A 52 inch HDTV screen can be found for as little as $1800. Some of the better HDTV manufacturers are Samsung, Sony, LG, and VIZIO. You really have to decide for yourself in the store when you look at the picture quality and cost of the system.

If you have ever wanted season tickets than you must have an HDTV set watch full screen high definition sports. There really is nothing like it in technology today. You can actually save money on game tickets by buying a big screen HDTV. You will also be able to create your own home theater environment for movies in HDTV. The full format is sure to please even the most critical TV viewers.

Today you can get a full sports package from the satellite company to watch ever game in the NFL on any Sunday. FIOS has the same option for the MLB and NBA. These channels are located in the 1900 channels near the end before music channels. The extra cost may or may not be worth the investment to you. It is nice to be able to see any ballgame in America at any time. This is truly a revolution in sports viewing for the entire country and not just the Northeast. The nice thing is that I can watch the Yankee network, the Phillies, or the Orioles and Nationals closer to home. This is extremely convenient compared to going to every game.

## 13. Traffic Devices

My research into major city traffic device networks led me to this great webpage below.  The webpage has links to every major city traffic camera in the world. Some of the links are old and do not work but if you try a few you may be pleasantly surprised as I was.  You may have to understand Spanish, Japanese, Chinese, or some other languages but the images are the same all over the world. Traffic jams are traffic jams.  The London website was run by BBC and is called "JamCam".  It only takes still images.  Images are time stamped with the local time.  The Madrid website takes still images also and is done in Spanish.  Caltrans is an interesting website of still images.  Maryland has full video and in this regard is a leader.  Very few of the websites are full streaming video.

Figure 12.  World City Traffic website

www.ecoplan.org/kyoto/challenge/traffic.htm

Ponder
World City
Traffic --
On-Line

---

- Useful downloads
- World Cities Inventory
- Thanks
- Click to add your city's traffic cam here?

It has long been our contention that one of the critical things holding us back from creating better, fairer and more sustainable transportation systems relates to our collective inability to imagine the cities that we would like to live in. It s, in fact, a double bind. We are, to try a phrase, literally blinded by the present as we look to the future.

To this end one of the main themes behind the New Mobility approach is that we need to learn to look more clearly at what we have, to see it for what it is, and then to see if we can move beyond that to something that is better and more sustainable. The several hundred views of traffic around the world that you have here provides food for thought. Here you will find a selection of real time views of traffic on city streets which provide some pretty interesting one-click coverage of how things look today in a huge variety of

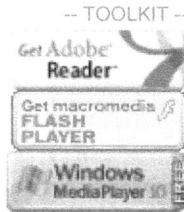

settings around the world. (Note: Some of these are official traffic sites, some with streaming real time video, others just cams that someone has aimed at a street. That too offers pause for reflection.)

If you spend a bit of time pondering these images, including at different times of the day and on different days of the week, several rather interesting things may jump out at you. For example, in city after city, country after country, how few hours of the day all these expensive roads are in fact being used to anything approaching capacity (and clearly beyond if you look closely). Hmm. That is worth at least a passing thought in our present context here.

And if only you could zoom in on all those cars on the street during the peak periods, you would see that in most of them there is just one driver, taking up something on the order of one hundred square meters of scarce taxpayer funded city real estate (by the time you factor in reasonable spacing, parking, etc.). Hmm,

Which brings us to the following question which is at the heart of this entire cooperative effort: Do we need to keep on building roads to solve our mobility problems? Or to get better at using the huge infrastructures that we have already put into place at enormous cost to the taxpayer.

## World Traffic Camera Inventory:

*(It's still the Net and work in progress, so from time to time some of these sites can go badly wrong. So if you get nothing in one place, keep moving. And let us know if you encounter a problem. Perhaps we will be able to do something about it)*

1. Aalborg, Denmark
2. Anchorage, Alaska, USA
3. Andora
4. Amsterdam, The Netherlands
5. Barcelona, Spain
6. Beirut, Lebanon
7. Belfast, North Ireland
8. Belgrade, Serbia
9. Bilbao, Spain *(in own window)*
10. Boston, MA, USA
11. Bratislava, Slovakia
12. Brno, Rep. Czech
13. Buenos Aires, Argentina
14. Chicago, IL, USA

15. Copenhagen, Denmark
16. Central Israel
17. Donetsk, Ukraine
18. Durban, South Africa
19. Dubai, UAE
20. Dublin
21. Ekatherinburg, Russia
22. Edinburgh, Scotland
23. Florence, Italy
24. Geneva, Switzerland
25. Glasgow, Scotland
26. Gothenburg, Sweden
27. Groningen, the Netherlands
28. Helsinki and region, Finland
29. Hobart, Australia
30. Hong Kong, China
31. Hong Kong 2, China
32. Honolulu, Hawaii, USA
33. Houston, TX, USA
34. Istanbul, Turkey
35. Karachi, Pakistan
36. Kiev, Ukraine
37. Kuala Lumpur, Maylasia
38. Lapland, Finland
39. Ljubjana, Slovenia
40. London, U.K.
41. Lucerne, Switzerland
42. Macao
43. Madrid, Spain
44. Manila, Philippines
45. Maryland
46. Mexico City, Mexico
47. Mexico City 2, Mexico
48. Milan, Italy
49. Montreal
50. Moscow, Russia
51. Nantes, France
52. Nashville, TN, USA
53. New Orleans, LA, USA
54. New York City (streaming video)
55. Nottingham, U.K.
56. Nowy Targ, Poland
57. Ontario, Canada
58. Osaka, Japan
59. Ottawa, Canada
60. Paris, France (streaming video)
61. Perth, Australia
62. Philadelphia, PA, USA
63. Prague, Czech Republic
64. Raleigh, N.C., USA
65. Riga, Latvia
66. Rijeka, Croatia
67. Rome, Italy
68. Saint Petersburg, Russia
69. San Sebastián-Donostia, Spain

70. Santiago
71. Sao Paulo (streaming video)
72. Seattle
73. Seoul, Korea
74. Stockholm, Sweden
75. Szegeden
76. Taipei
77. Tallinn, Estonia
78. Teheran, Iran
79. Tokyo, Japan (real time)
80. Toronto, Canada
81. Trnava, Slovakia
82. Turin, Italy
83. Vyborg, Russia
84. Vienna 1, Austria
85. Vienna 2, Austria
86. Vilnius, Lithuania
87. Vilnius, Lithuania (2)
88. Warsaw, Poland
89. Washington State
90. Weillington, New Zealand
91. Wuppertal, Germany
92. Zurich, Switzerland

The following cameras are extracted from the NYC DOT website. They have still images when you click on the traffic camera you desire to look at. The size of the traffic problem in NYC is huge. This makes the solution noteworthy.

Figure 13.  NYC - 5 Burroughs Traffic Camera Networks

The Bronx

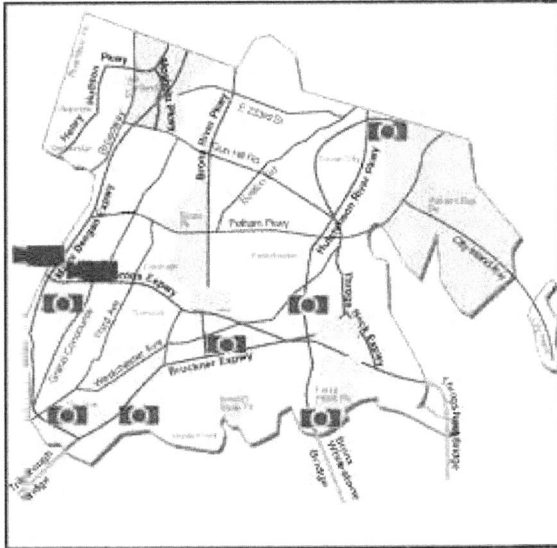

Queens

Manhatten

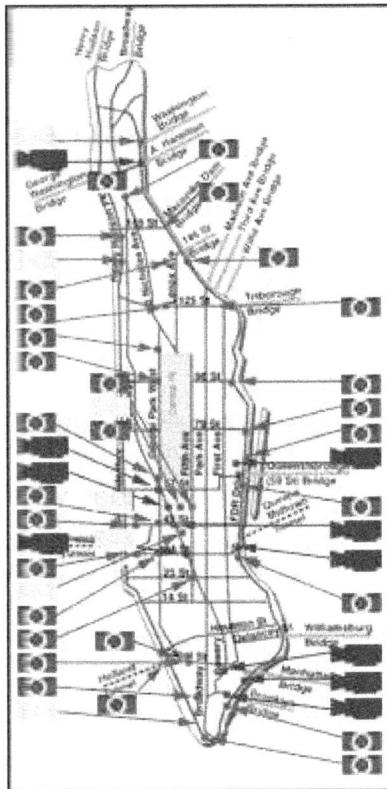

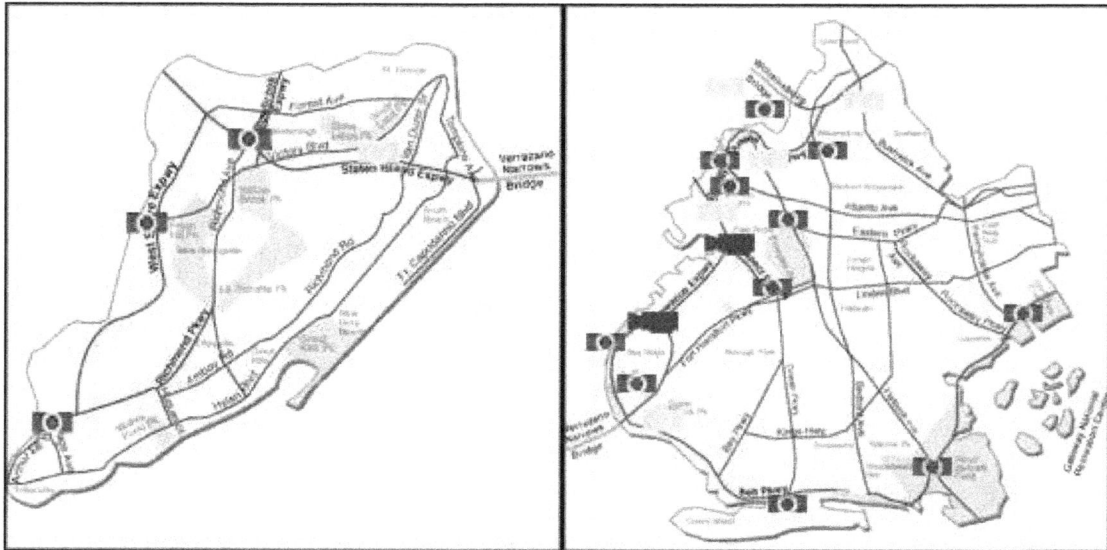

Staten Island                    Brooklyn

Red Light Running cameras are a subject of much controversy with the public. These cameras are not networked and stand alone at intersections. They are monitored and film is downloaded by contractors. The estimated cost of a traffic red light camera was $50,000 in 2005. There are basically two types. A wet film type and a dry film type. The photo is triggered by the vehicle in the view of the camera after the light has turned red. Many times a vehicle may be stuck in the intersection behind another vehicle and not know it until they receive a picture in violation. Some judges will throw these pictures out of court but for the most part the legal pictures of violators are upheld by a wide margin (97%). The evidence in in the photo itself. Most of them show the traffic light as red and the vehicle in the intersection. Interestingly enough, the cost to pay an officer to watch an intersection is fare greater than the cost of a red light camera. Thus the economic benefit outweighs having an officer watch an intersection. The safety benefit is that people will not violate again if they have paid a fine. It is a deterrent to future violations and thus the traffic is safer. Red Light cameras are not revenue generators even though they can bring in quit a large sum of money. And today security video cameras are everywhere so there is no escaping "big brother".

By law red light cameras seem to be curbing speeding violations but the revenues fall off after the drivers realize there is a speeding camera at an intersection. This is creating more safety. It is not clear whether the majority of the drivers just speed up after the red light camera or not. This is called "manipulating" the red light camera[20].

---

[20] Maryland SHA Daily News on internet

## 14. Railway Systems

The AMTRAK runs through the northeast corridor from north to south and east to west. Rail travel is not as big in the states as it is overseas. It could be more significant if more people rode the train. Commuter trains can be expensive for monthly tickets. And you have to be on time every time. The fact that you can read a paper while you travel to work is nice. The major stations are Union Station, Camden Yards, and Central Station. The rail system is older than the highway system. It is very reliable. If I had to choose between a bus and a train, the train wins every time. The trains in large cities in the northeast are integrated with the subway systems. That is, the subways also run from the stations where the AMTRAK connector runs. This is important for the visitor to the city who needs to move around the city once he arrives. The fees are somewhat better than cab fares. On the longer trips you can take advantage of the club car and dining car of the train. There is also the sleeper births for overnight train travel. The airlines have put trains out of business in some respects but they are still a nice way to travel. Train systems include a schedule system and network of telecommunications between stations. This dates back to the telegraph which controlled departure and arrival times of early trains.

Figure 14. World Trade Center Subway Station

Diesel electric trains run the cargo trains. These can be long freight trains with miles and miles of cars. They may run on the same tracks as passenger trains. They cost significantly less than other modes of transport. Trucks can interface with cargo on freight trains at the freight yards. The economy of having a rail network in place can also be very secure if managed properly.

There is a 5 minute clock on finding very freight car in every train network in America. This was placed on each car by Homeland Security to make sure they could find any type of cargo at any time. A RFID locator can be used to tag each freight car on the rail network. It will have a GPS locator and manifesto. This is a powerful deterrent to crimes and terrorism on the railways. The most concern here is for the chemical rail cars carrying hazmat materials. They are certain targets for terrorists. Consider that freight trains in America are routed through high and low population areas and you have a moving time bomb. But rail companies have people watching the rails. It is not a simple matter of sabotage any more. We also have people watching for videotaping and cameras used at freight train locations. The involved citizen may be a good deterrent.

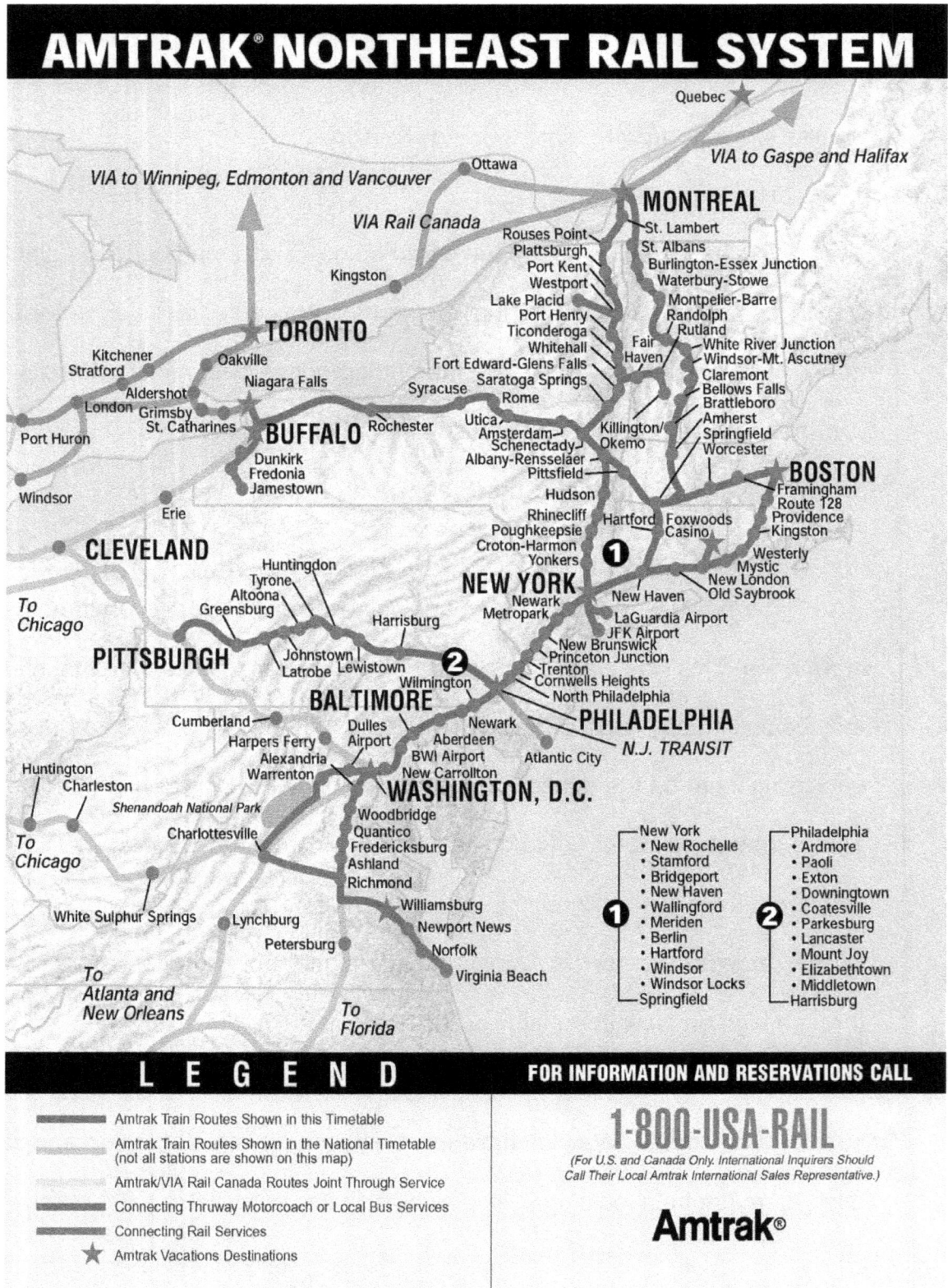

Figure 15. AMTRAK Northeast Rail System

## 15. Bus Systems

New buses have air brakes, air shocks, and air conditioning and they are a pleasure to ride in. Public buses are generally this type. Any time you can get a charter type bus that is good too. The bus is economical and saves you the headache of driving.

Scheduling and routing of buses is most important to let riders know when they will be at the bus stop and what they will cost. A good bus network has buses running from one city to another. Buses run in the CBD and to the suburbs. These are regular routes held to a tight time schedule.

Integration with rail and airports is also important. A good bus line will stop at railway stations and airports to ensure passengers can transfer to another modal. The times are very critical when you have to make the next train or airplane. You absolutely have to be on time.

Park and rides exist at many locations where you can interface with your personal vehicle. The bus may stop there and take on many passengers for a commute to the city. Usually Park and rides are located in more rural areas.

Economic ways to ride the bus include staying on until you come back to the stop where you got onto the bus. You can also ride a loop and a half to your destination if you miss your stop. This will take more time but you will get there.

Hawaii has a bus that is cheap for example - 60 cents will take you anywhere on the island of Oahu. This type of city bus line is welcome to the walker or non- car owner. In tourist cities like Honolulu this is welcome.

In the northeast corridor buses go from one city to another all the time. Washington bus terminals take tickets for Baltimore and Columbia. The Route 29 Flyer goes from Columbia to Washington DC at K Street. This is typical of how to traverse from one city to the next. There are also Greyhound terminals in the major urban areas. They take people long haul from one city say New York to another such as Baltimore. This is very convenient and inexpensive for the rider.

Figure 16. I-95 New Jersey

Figure 17. I-95 DelMarVa

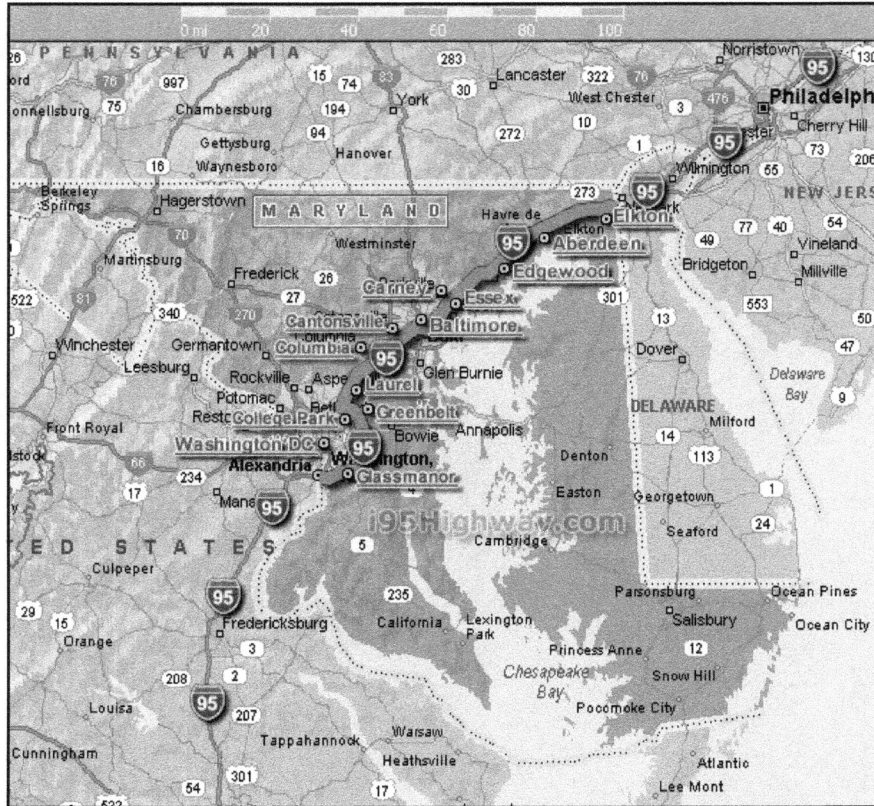

16. Airport Systems

The northeast air corridor is quit crowded. There are 35 airports in Maryland alone not counting the smaller private fields. The Federal Aviation Regulation (FAR) is the regulating authority for public commercial aircraft. The Air Force Regulations are the authority for the Air Force. Naval Regulations are the authority for NAVAIR. The landing strips are configured with radar and communications systems. These are complex. The FAA has been updating the Air Traffic Controller System for many years now and is looking toward the NEXTGEN system. One of the major problems with the current system is that it is not user friendly. The older system was also based on mainframe computers which are becoming more obsolete. The FAA does interact with the military on flights in the airspace adjoining each other. One of the greatest problems of 9/11 was the lack of immediate coordination of the FAA and military on a response to the attacks of the hijacked commercial airliners. Planners did not think outside the box enough to predict that an attack on several airplanes could occur at the same time.

Figure 18. JFK Airport

Every plane has computers on board that basically fly the plane. The pilot is a systems analyst when it comes to managing the computer systems that control flight surfaces and flight systems. Anyone can learn some of these systems through time spent on flight simulators. However, only real time real piloting can achieve the flight hours required to master the flight of major large aircraft. Command pilots have over 5,000 hours in the cockpit. They also have to be in good physical shape to fly. The physiology of flight is such that it stresses the body to limits not known on the ground. In this regard flight systems include human physiological systems. There is a lot of human factors engineering in flight controls.

Civil engineers need to know aircraft performance in order to build runways and hangers. Flight engineers need to know how to fly and how the aircraft performs under stress. The 9/11 Commission Report suggested that we need to protect airports better. Transportation Safety Administration (TSA) has this mission. They are a part of DHS. Airport security is not easy to implement but it is very necessary under current threats and vulnerabilities. 9/11 hijackers could have been prevented had flight instruction schools been on the lookout for middle eastern men who did not learn how to land an airplane. The passengers need to be separated from the pilots by a locked doorway and the pilots need to be armed to ensure security or an air marshal needs to be aboard the plane. The systems used to screen passengers are getting more sophisticated and can detect gels and liquid explosives. This protects the in flight integrity of the aircraft.

Airport runways have a certain distance they have to maintain for different aircraft sizes. If a runway is too short the aircraft can overshoot the runway. Various large runway airports have to be available for emergency situations in the region to take diverted air traffic and land it safely. In this regard all the systems are interconnected. Also each airport has an invisible cone around the airport for landing aircraft patterns. This is outlined in the FAR. Different flights travel at different altitudes. But the airport is where they reach down to land safely.

Each airplane has a transponder in it that beacons the location to the ATC system. Newer GPS technologies also make sense in this environment. The GPS location is unique at every location on the planet. Software engineers ensure that the software that runs the plane is usually highly reliable ADA source code by AONIX or some other company. The Air Force Software Technology Center[21] (STC) at Ogden Utah develops source code for aircraft for the Air Force under tight requirements. Any errors in the software are debugged in simulations before actual flight. However there have been issues with some aircraft like the F-22 whose software did not perform correctly in flight tests. It is extremely hard to if software will perform without flight testing. Since most modern aircraft are

---

[21] STC ct Hill AFB, Ogden Utah. Works with Utah State University and SEI at Carnegie Mellon University. Crosstalk magazine is one of their products to the software development world.

fly-by-wire this is critical. The aircraft systems need to be secured in a way that disallows intruders during flight operations. The C programming language has been used for satellite computers. The best programmers would be those who could also fly an aircraft.

Nav Aids are known as navigation aides to flight. These give the pilot the location and direction of flight. The attitude and altitude of the aircraft is also indicated. The  airspeed indicator gives flight speed. The engine RPM tells the status of the engines. The NAV radio frequency talks with the tower and ground control. The radar system tells where other aircraft are flying in relation to your aircraft. These systems have to work in every aircraft all around the airport.

Inside the airport there is the flight destination/arrival schedule computer. This board posts the expected times of departure and arrivals. The airline reservation systems take passenger reservations and book seats on individual airlines via computer servers. The customer uses his computer to sign up and pay for the flights. This modern system has been achieved with Personal Computers. The predecessor systems were archaic on mainframe batch systems. Interactive computing has brought real time reservations to the flight industry.

Figure 19. NEXTGEN Traffic Control System

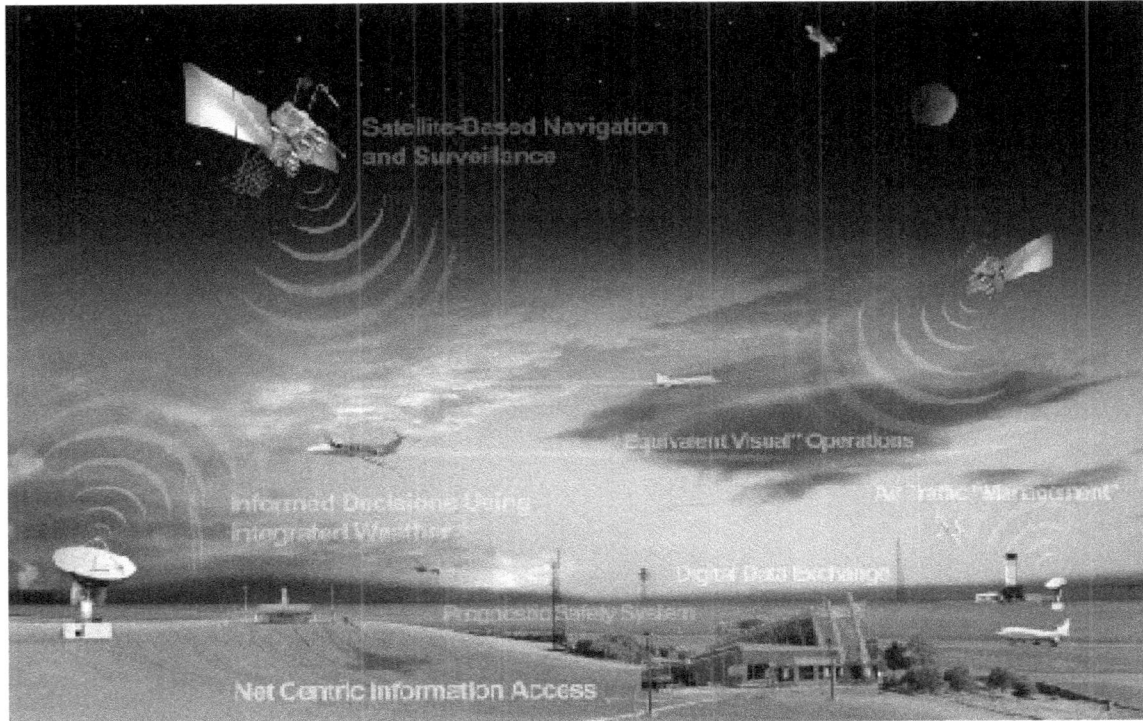

NextGen will have the technology and infrastructure necessary to handle the increasing air traffic expected in the coming decades. Satellite-based navigation will allow aircraft to fly more direct routes and navigate around inclement weather which will increase airspace capacity and reduce delays.

Furthermore, the FAA webpage states that

"The FAA is working on a plan now to make the best use of new and existing technology, infrastructure, and employees to handle the doubling and tripling of air traffic expected in the coming decades. The Next Generation Air Transportation System, or NextGen, will transform the national airspace system from one that is based on ground radars to one that uses satellite technology.

**As FAA moves through the transition of air traffic control from radar-based to satellite-based,** controllers and pilots will have the advantage of increasingly sophisticated technology. While radars require aircraft to fly over their physical locations on the ground, satellite-technology will allow controllers to guide aircraft in more direct routes through the nation's airspace. Essentially, every controller in the United States will be able to see the exact position of every aircraft flying in our airspace, no matter where in the country they work."

17. Seaport Systems

The major ports in the northeast are New York Harbor, Fall River, Boston Harbor, Baltimore Harbor, Philadelphia, Wilmington, and Washington DC. These all vary in size but they all function similarly. The largest one is probably New York Harbor which extends up the Hudson and East Rivers. Boston Harbor is the most historic famous for the Boston Tea Party and home to the USS Constitution. Baltimore is close behind with major historic events during the War of 1812 at Fort McHenry near the entrance to the harbor. The USS Constellation is moored in Baltimore harbor. Philadelphia port is on the Delaware River. The USS Olympia used to be moored there. The Fall River is the home of the USS Massachusetts. Wilmington is on the Delaware River also. You can tour these places by car on I-95. This was one of my first vacation site seeing trips when I worked for the Navy. My father also did radar and sonar work at Groton and New London Connecticut sub bases for the Navy. He worked at Brooklyn Navy Yard when he was a teenage welder ship fitter at Bethlehem Steel before he entered the Air Force in 1952.

Water Navigation systems are required in any busy port. The ships must know where they are going to dock and what channels are clear for approach and docking. These systems are controlled by the bridge of the ships and the ports authority main engineering building. The IT system called NAVIS is the one used at the port of Baltimore. The US Coast Guard also uses navigation systems. Sonar is used to scan the bottom of the water body. In Baltimore the Chesapeake channel runs 39 feet deep at the centerline. Ships entering and leaving the port follow the channel. When the ship reaches the Bay bridge it is instructed to stay under the main span of the bridge.

Figure 20. Port of New York

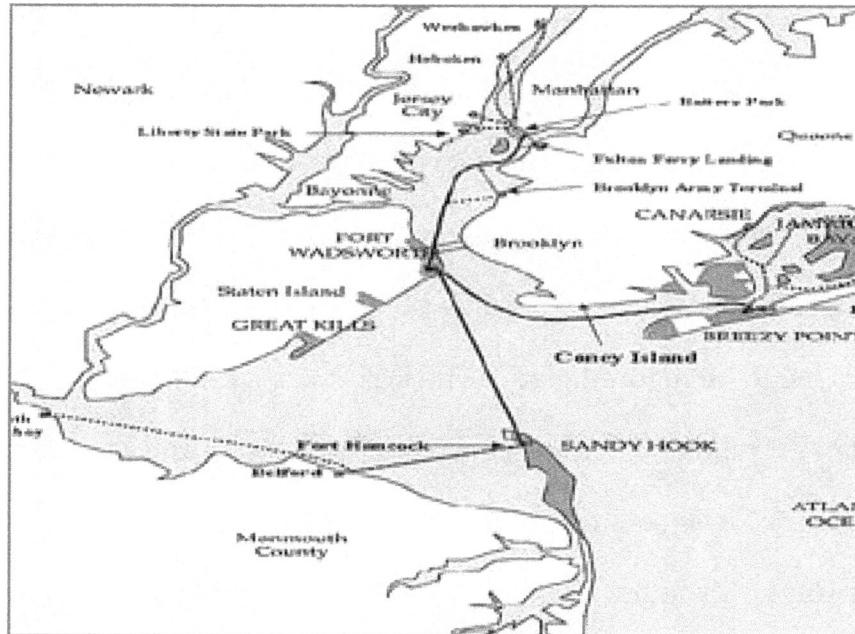

The manifest tracks passengers on cruise ships and cargo on cargo ships. The electronic manifest is key in helping identify passengers and cargo and keeping an updated list. Sometimes cargo can be also identified with RFID tags.

Cargo Ships are loaded to maximum capacity with sealed containers. These containers are tracked from port of departure to port of entry. For homeland security purposes containers are scanned by radiologic scanners at the port of departure for nuclear materials onboard. This program has increased since 9/11. Sealed containers are transferred to trucks and rail cars when they are offloaded at the port. Additional scanning for nuclear materials can be done at weigh stations and railroad yards. The huge cranes at seaports offload the cargo containers and these are massive sized steel structures designed to carry tons of cargo containers. These systems are operated by longshoremen who work at the port docks. This requires a unique skill set to operate a container ship crane.

ICCTV Video cameras can verify the path ships take and a network of marine cameras is being developed in Maryland to track ships on the Chesapeake. The system when completed will link all of the cameras on the shoreline with those on bridges and high points. If there is an emergency officials tracking the vessels will know immediately.

Radar is used to identify surface objects such as planes and other ships. The signals return from the craft after hitting the surface and return a signal strength the size and amplitude of the object. Military radar systems are very sophisticated and can be linked to weapons firing control systems for defensive measures.

Radio and Satellite Communications systems work on the sea by bouncing signals off the ionosphere and satellites. These are much more sophisticated than they were many years ago. GPS locators also pinpoint the location of ships at sea very accurately.

Propulsion systems power the ships. They range from diesel to nuclear propulsion.  Air jets can be a propulsion system on some newer Navy ships and cruise ships.  Propulsion is critical and needed in critical moments of the voyage.

Figure 21.  Islands of New York

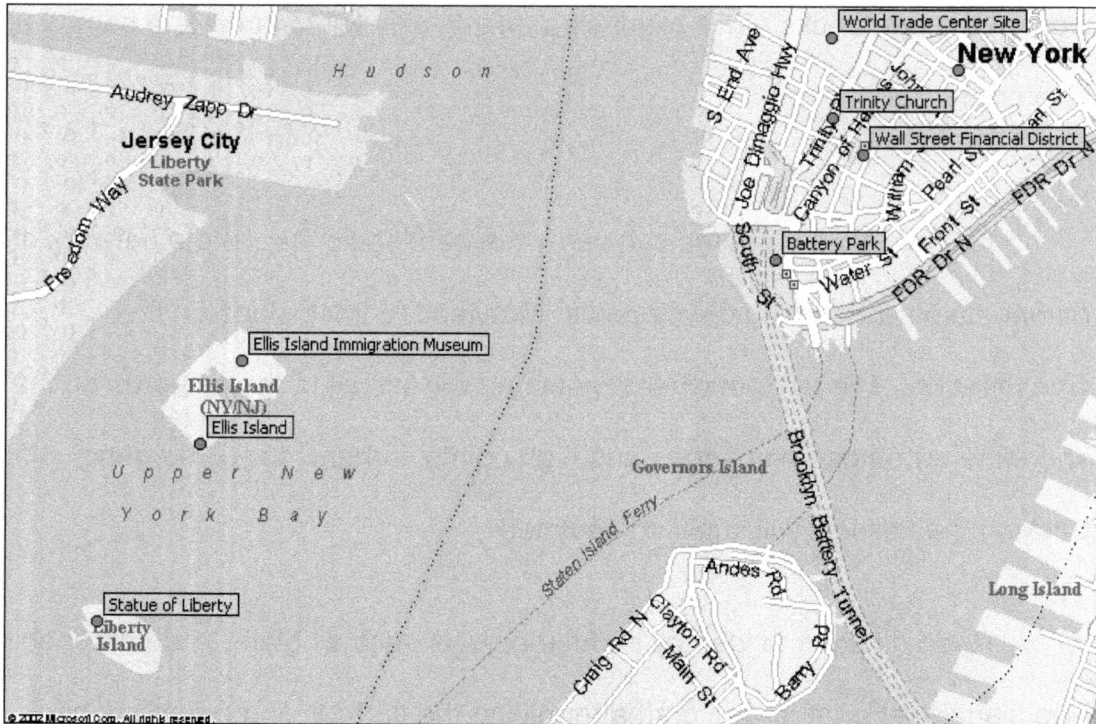

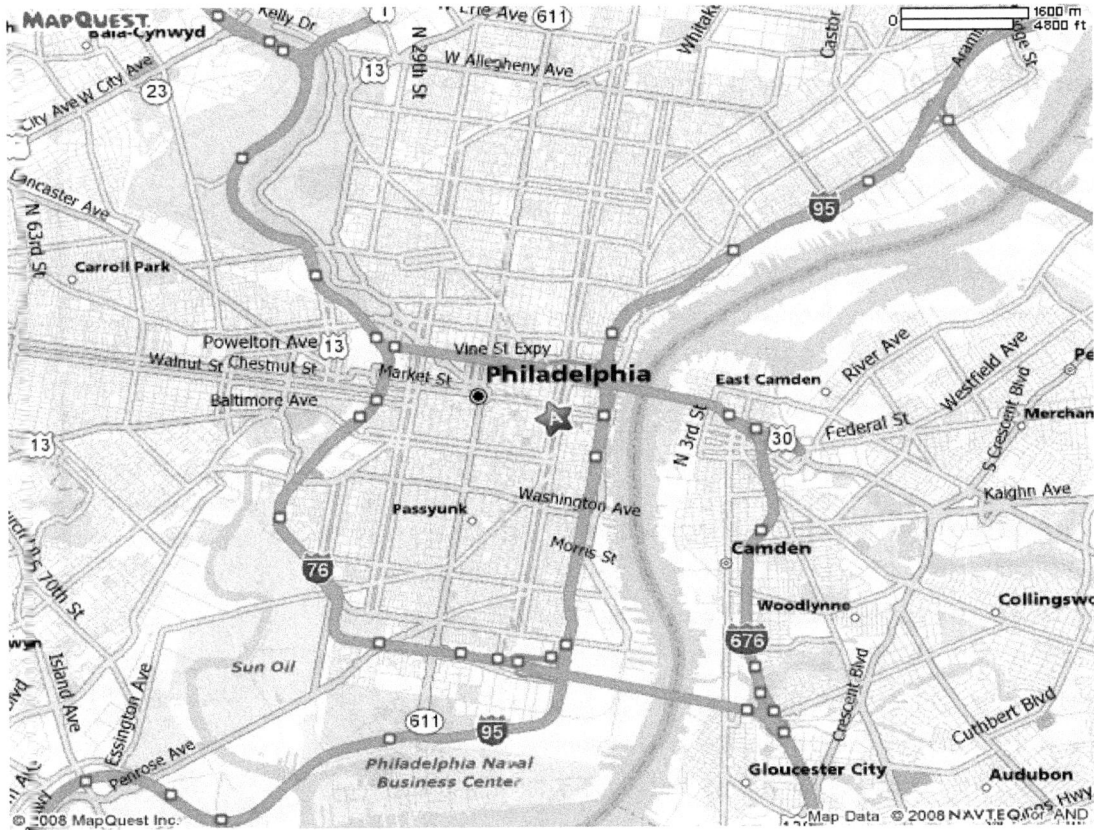

Figure 22. Port of Philadelphia

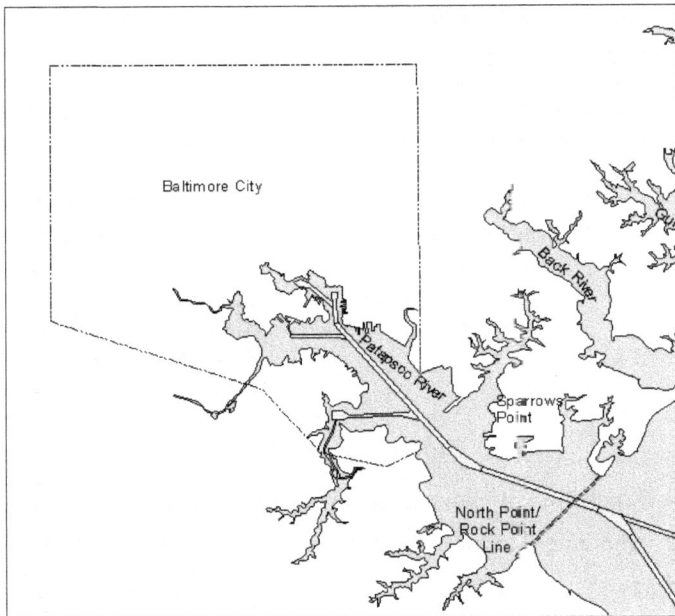

Figure 23. Port of Baltimore

## History of the Port Authority of New York[22]

### The Beginning

The impetus for the Port Authority's formation can be traced back more than 300 years. First, there was the accident of political history that divided a common port area between what ultimately became the states of New York and New Jersey. In time, the division of the harbor - a vital source of commerce and growth - led to controversy in the region.

An aerial view of the New York-New Jersey Harbor, a vital source for regional commerce and growth.

Throughout the 19th century, New York and New Jersey waged many disputes over their valuable, shared harbor and waterways. A dispute over the boundary line through the harbor and the Hudson River - settled by the Treaty of 1834 - once led state police to exchange shots in the middle of the river. The impasse eased when the two states agreed that the port area was, in effect, one community and that conflict squandered the port's potential. The states sought a governmental body to oversee port affairs and found a model in the Port of London, administered by what was then the only public authority in the world. On April 30, 1921, The Port of New York Authority was established to administer the common harbor interests of New York and New Jersey. The first of its kind in the Western Hemisphere, the organization was created under a clause of the Constitution permitting Compacts between states, with Congressional consent. An area of jurisdiction called the "Port District," a bistate region of about 1,500 square miles centered on the Statue of Liberty, was established. In 1972, the organization's name was changed to The Port Authority of New York and New Jersey to more accurately identify our role as a bistate agency.

Visionaries on both sides of the Hudson River join New York and New Jersey's shores in the spirit of partnership and growth.

### Decades of Experience

In 1930, the two states gave the Port Authority control of the recently opened Holland Tunnel as a financial cornerstone. Immediately, we began blazing new paths in transportation, engineering, law and administration - a precedent that is still seen today through our commitment to the people of this region, and the region's advancement. The Port Authority's first charge was to construct the critical interstate crossings in the late 1920s and early 1930s, including the George Washington Bridge, Outerbridge Crossing, the Goethals and Bayonne bridges. In 1937, the first tube of the Lincoln Tunnel was completed.

# Urban Systems Engineering

In the late 1940s, at the cities' request, the Port Authority leased three airports from Newark, NJ and New York City in anticipation of the jet age. Newark and LaGuardia airports, along with an infant airport on a large meadow destined to become John F. Kennedy International, were linked into a regional aviation network. In the 1950s and 1960s, we built the Port Authority Bus Terminal and added a second deck to the George Washington Bridge. We completed the Lincoln Tunnel's third tube, rebuilt many Brooklyn piers and developed the world's first containerports at Port Newark and the Elizabeth-Port Authority Marine Terminal. The Port Authority also acquired the Hudson and Manhattan Railroad and began operating it as the PATH rail transit system.

In the 1970s, we helped advance the region's interests in port and trade promotion through construction of The World Trade Center, which brought together private firms and government agencies engaged in international business. The two towers became the hallmark of the New York City skyline, reaching higher than any other skyscraper in the city. Today, the memory of the Twin Towers remains a symbol of unity, freedom and strength as we help to rebuild Lower Manhattan.

During the 1980s, the Port Authority began a bistate program of industrial redevelopment, aimed at helping the region retain manufacturing firms and jobs. We developed various projects in the two states: The Teleport, a telecommunications center in Staten Island, NY; the Legal Center in Newark, NJ; a resource recovery plant; industrial parks and waterfront development in New York and New Jersey.

In the mid-1990s, we concentrated our efforts on the transportation and trade projects that constitute our core mission. Among significant improvements, we developed ExpressRail, the on-dock ship-to-rail transfer terminal at the Elizabeth-Port Authority Marine Terminal; enhanced facilities at our three regional airports; installed a system to provide train status information in stations to PATH riders; and introduced the E-ZPassSM electronic toll collection system at our tunnels and bridges and an Intelligent Transportation System at the George Washington Bridge.

18. Bridge and Tunnel Systems

This chapter shows some of the bridges in the northeast along the I-95 corridor. We also show how network analysis can be used to determine risk of attack on a bridge. A bridge or tunnel is a high value target to terrorists. We have caught some people videotaping operations at our bridges in Maryland since 9/11. The main purpose is to explain how these assets form a network on top of the interstate highway network. We could not run the interstate system effectively without these bridges and tunnels.

Boston Tunnels and Bridges

The Big Dig in Boston was a project to fix the tunnels near Logan International Airport. The ceiling of the tunnels collapsed and killed some motorists. The project hoped to fix the problem and open to the public. Boston harbor is the water body under which the tunnels reside. The highway cuts through downtown Boston. I-95 is actually many miles away from the city to the west.

New York Tunnels and Bridges

There are many bridges in New York from each of the islands connecting to Manhatten. The most famous is the Brooklyn Bridge. The Battery Tunnel connects Manhatten to Brooklyn. The Holland and Lincoln tunnels connect Manhatten to New Jersey. Highway traffic can be terrible in New York and there are always toll fees wherever you go. The best way is to map out a route and take someone who knows the way. At least this has worked for me over the years. I get one of my cousins to drive me where I need to go. Some of the other bridges are the Verrazano Narrows Bridge from Staten Island to Brooklyn and the Outerbridge Crossing from Staten Island to New Jersey. These are relatively newer bridges in the history of New York. The George Washington Bridge is located north of Manhatten on I-95. It connects upper NYC to New Jersey. The wait time can be great at the GW Bridge.

The pictures below show some of these bridges and tunnels. They form a network of bridges along the New York freeway network system. Each bridge is a node on the network. In network analysis[23] we can determine the risk of any given node (bridge) on the network to terrorism. This risk analysis uses the fact that the bridges are primary targets. We rank the bridges and then spend money to protect the primary bridges which are the most valuable. The rank order of the bridges is important. The George Washington bridge standouts because of the amount of traffic and the location on I-95. Shutting down this bridge brings traffic to a north south standstill. The traffic is already very congested at the GW bridge depending in the time of day and week and local events.

Figure 24. George Washington Bridge

Figure 25. Holland Tunnel

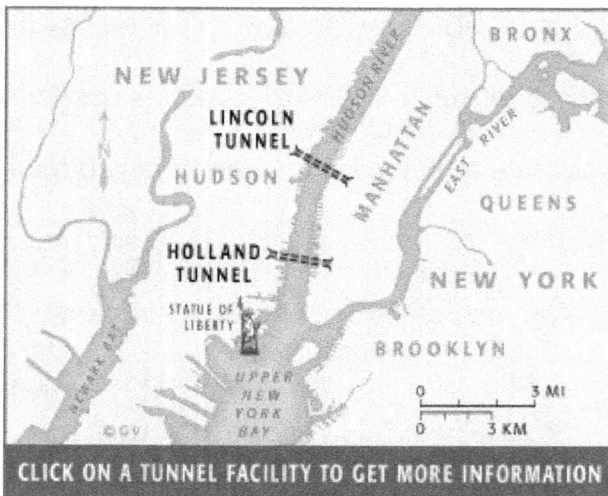

Figure 26. Tunnel Locations

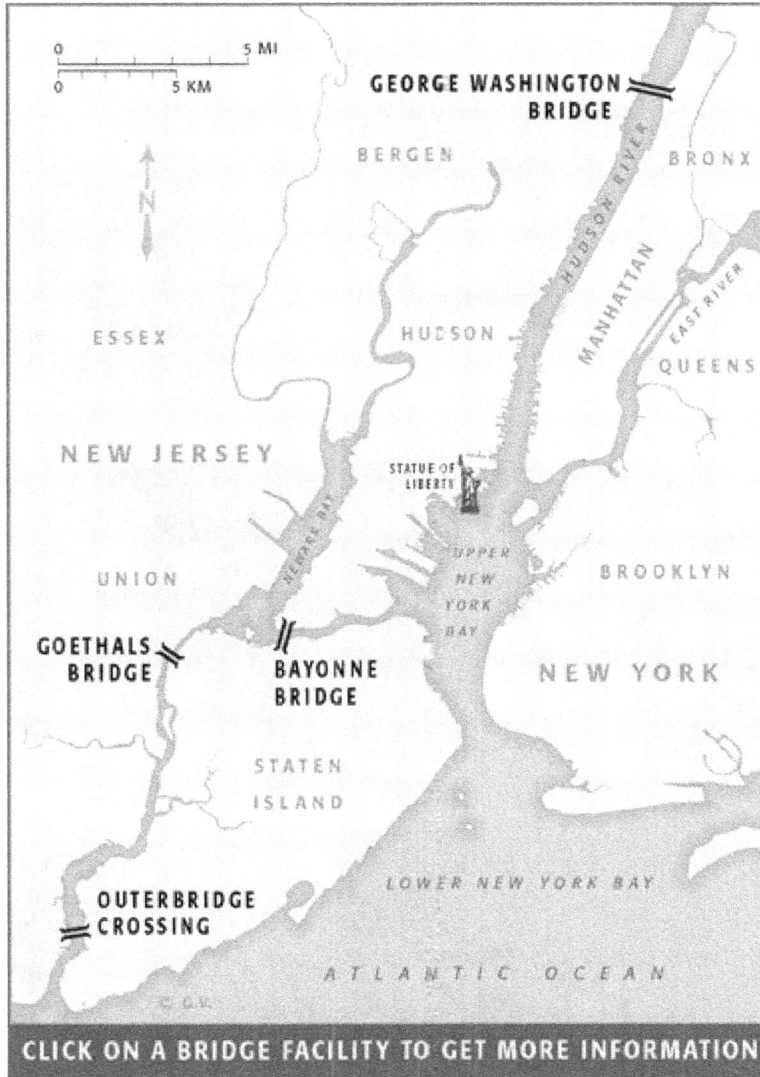

Figure 27. Bridge Locations

Figure 28. NYC Bridges

Outerbridge Crossing      Goethels Bridge      Bayonne Bridge

Manhatten Bridge

Whitestone Bridge

Throgs Neck Bridge

Brooklyn Bridge

Triborough Bridge

Verrazano Bridge

Williamsburg Bridge

Baltimore Tunnels and Bridges

Baltimore has two tunnels – the Harbor Tunnel and the Fort McHenry Tunnel. Both are located on I-95 where it crosses the harbor. The bridge near Baltimore is called the Francis Scott Key Bridge located on I-695 around the beltway on the Chesapeake Bay. The Bay Bridge is located on Rote 50 in Annapolis. Baltimore Harbor Tunnel was built in the late 1950's. The Bay Bridge was built in 1964. The second span of the Bay Bridge was completed later. Travel north on I-95 must go through one of the tunnels or bridges. The alternative is the I-695 beltway around Towson. The State Highway Administration built all of these projects. The State Operations Center has video cameras at each of the bridges and tunnels and can view operations.

Figure 29. Chesapeake Bay Bridge

Delaware Memorial Bridge

The Delaware Memorial Bridge sits on I-95 and connects New Jersey and Delaware. The access point to the New Jersey Turnpike is north of the bridge.

Figure 30. Delaware Memorial Bridge

Washington DC Bridges

Washington has plenty of bridges but no tunnels. The marsh ground will not support tunnels around Washington. At one time gravel and dirt had to be carried in to fill in downtown Washington around the Jefferson Memorial. The bridges are all designed to be draw bridges and gothic. There are no suspension bridges here. The 14th Street Bridge, The Roosevelt Bridge, The American Legion Bridge and the Woodrow Wilson are all find bridges but none are state of the art. The Woodrow Wilson Bridge was recently rebuilt. DCDOT runs the traffic cameras in the city and they can track vehicles across the city.

Figure 31. Washington DC Bridges

Woodrow Wilson Bridge

Theodore Roosevelt Bridge

19. Intelligent Transportation Systems

CVO means Commercial Vehicle Operations and include all trucks, buses, and other commercial vehicles. Commercial vehicle traffic is a subset of all traffic. CVO has an impact on the economy and large sums of money are invested in commercial vehicle transportation. The CVO is managed at the state level by Motor Carrier Divisions in state transportation departments. The commercial vehicle associations help the trucker companies organize for the benefit of all. The FMCSA manages all trucking from the federal level. The FMCSA Regulations are the rules of the road for commercial vehicles.

CVISN means Commercial Vehicle Integrated System Network and it is a group of software networks that manage commercial vehicles. There are several websites like MCMIS and SAFER. There are transponders in trucks that allow for Weigh in Motion (WIM) machines to weigh trucks while they are in operation. These may be substituted for full blown weigh stations at a reduced cost.

ICCTV means Integrated Close Circuit Television. These systems of video cameras are used for security and other purposes. Cameras may be placed at certain points around the highway system to monitor traffic. Monitored traffic allows emergency vehicles to arrive at the scene of an accident on time. License Plate Readers may also be placed around the system for security.

RFID means Radio Frequency Identification and is used on sealed containers and other loads of freight to ensure security. An RFID locator can tell where the container is located. It is powerful in that any freight can be located in 5 minutes. This is a national goal of all freight carriers.

In Truck Computers hold scheduling information. The trucks also have portable PCs and laptops available to them to help with managing the trips.

Automated Weigh Stations are weigh stations with advanced computer systems for weighing loads. The computers are hooked to load cells or plates that weigh the trucks as they drive up to the window of the weigh station. The read out on the computer tells how much the load weighs and whether it is over or under weight. If the load is overweight the officer cites the driver and a fine is paid. Trucks may also be pulled aside and inspected for mechanical problems. When a truck is in violation of repair it is pulled out of service until the repairs are made. Many companies try to run older trucks and avoid maintenance costs. This only makes them more eligible for out of service time at the weigh stations. Some types of problems are bad brakes or other brake problems. Any hydraulic fluid leaks are noted. Other types of violations are listed in the Maryland Vehicle Law handbook.

Maryland SOC or State Operations Center is the control center for state transportation operations. Video feeds connect the state highways to the control room. Computer operators look for trouble and react with a phone call to State Troopers and highway personnel to deal with the problems on the highways. Variable message signs can relay alternate routes and information to drivers on the highway when a backup is noticed on the video feeds.. Highway advisory radio stations can play messages that are sent from the SOC to warn motorists of impending dangers and backups. This is the primary idea of managing the highway congestion with electronics.

Incident Management & Response is the act of obtaining help for motorists if there is a problem on the highway. The incident is recorded by the SOC and state troopers or highway emergency personnel respond within minutes to the site of the accident or incident. Required action is then taken to save lives and property.

20. Urban GIS Systems

A GIS system is a "Geographical Information System" and thus conveys information using maps and geography. Urban GIS systems are therefore systems that convey information using urban geography. The largest company in the GIS business is ARCVIEW and ESRI located here in Towson, Maryland. They are contracted to build the Statestat GIS software for Maryland. They also built the Baystat GIS. These software products are really coming of age. They show how money is being spent and where it is being spent on a map of the state. This hold people more accountable and gives a one shot snapshot of where problems may be. Although graphs and data are the highlight of these systems there are modules for GIS software.

GoogleEarth.com is a new mapping system that is part GIS part mapping software. It can go anywhere on earth and show you what it looks like from satellite. This is a powerful system. The only problem is that it is not real time. The cars and vehicles you see on streets are not captured at the current time. They are pre-recorded when Google earth takes the shot. Usually GIS system marry data and maps. Google is mostly a mapping software.

Corporate GIS system can have any type of data included in them. Maryland SHA GIS has 7 layers. A highway layer shows all the highways. An accident layer shows all the accidents on the highway. The end user navigates the layers and makes selections of data that are placed on the map. The county can be used as the map instead of the state. The city can also be used instead of state. The layers can hold any kind of data that is geographic in nature. An Oracle database is attached to the software to bring in the layers of data to the map. Oracle than displays the data on the map through the ARCVIEW interface. ARCVIEW is a very popular viewer for GIS applications. There is even a magazine called "ARCVIEW" by the company for end users.

Other types of GIS systems could include crime mapping in cities and suburbs. This type of system would map crimes already committed and be able to predict where future crimes may occur. A National crime map could also be develop for the FBI.

## 21. Designing a Network of Systems

Adhoc networks just pop up where there may be more than one node of information. The adhoc network is never planned but occurs as a result of adding more nodes to the network without really typing to build a network. There is no thought put into the design of the network.

Network of networks happen when multiple networks are linked together to form one giant network. Many types of networks connect together into a mesh of larger networks. The root network node is then desired to be known so that anyone can traverse the network.

Central Star network is based on a star pattern. The nodes come out of the star and connect back to the star node. Examples of this would be the solar system with the sun at the center of the system.

Distributed Star networks are based on the fact that there may be many central star networks inside a larger distributed star network. The central star may connect to a super star. This is like the many galaxies connecting to a super star.

Ring networks are formed in circles and usually a token runs along the network. If we are talking about transportation networks than we could be describing the highway system such as a Beltway around a city.

Distributed Rings can occur in one place such as the two Beltways in Maryland. They also connect and form a distributed ring of highways. Many LAN networks can connect and form distributed rings if they are Token Ring networks. The token passes from one node to the next until it completes a circle.

Bus Drop networks are straight forward in design. The are dropped off the ethernet or dumb terminal connections until there are plenty of connections. The Bus Drop also uses collision detection. This is the design for broadcast media.

Electronic and Civil networks collide in transportation networks with video cameras. The networks have co-existed for many years since the telegraph and railroad. Electronic devices can make transportation networks more efficient.

Economies of Scale impact the network since the more nodes the larger the network. The larger the network the more complex the troubleshooting. Also the more areas can be reached by the network. This we will call network coverage. Network coverage in urban cities is greater than rural areas but improvements are happening in rural areas to cover 100% of the landscape.

Network Ownerships in the Megalopolis are primarily public and private. The joint ownership means that the critical infrastructure is owned by both government and private agencies.  This has an impact in how we maintain and

patrol the networks for security. Government is usually tapped with protection of

networks.

22. Complexity in the Megalopolis

Electronic control is primary in the big city. Systems are controlled by a central control facility. The Ports of New York were controlled by the 102$^{nd}$ floor of the World Trade Center. You can believe that the terrorists knew this when they hit the towers. Complex systems are computer controlled and people watch the monitors 24/7. Nuclear power plants have similar SCADA control system setup on computers with monitor watchers.

Instant Access is available from any control point in the systems. Overrides can be plugged into the system to adjust power outputs. Shutdowns can be down from remote terminals in the networks in the city.

Cost Effectiveness is gained by the ability to manage all the resources marshaled through the electronic systems in the city controller. Reducing usage during high output periods and peak times is paramount to keeping maximum energy flow across the conduits.

Massive Interconnections are found between civil systems and electronic systems. The sewage system is large in NYC. It runs under the streets similar to the electrical power grid lines. Often they are located in the same trenches.

Rural resources feed the urban dweller in terms of food and soon they will also generate much of the power through wind and solar energy investments. Green buildings and smart buildings will take more energy and use it better  The return to the rural partner is the increase in automation that is available for rural tasks such as agriculture, recreation, exurbia, and religion. We need our rural areas for these parts of life that are crucial to living a good life.

CBD Controllers handle the bulk of the city network traffic.  Major population areas handle their jurisdictional concerns and leave others to be handled by other jurisdictions.  It is as simple as a line on a map for many managers. The meeting of the minds for neighbors is what is difficult because some solutions to problems need to be handled regionally and not by map line.

Figure 32.  Network Controllers

Network Controllers

Network Monitors

## 23. Rural Systems Upgrades

Full connectivity to Ex-Urbia is ensured by the interstate system. Anyone can reach from city suburbs to the rural farmlands in a matter of minutes. Many rural counties in the 60's are now suburban counties full of developments outside major cities. Travel to some major urban areas and you know this is true. I grew up in rural Topeka, Kansas and it is mostly farmland. A fun place for kids to grow up but hard work as an adult. The fact that farmers are having hard times now has led to them selling off properties to housing developers. Agriculture was once 90% of the economy before the civil war and now it is only 5% of the economy as the service sector has provided more jobs recently.

Rural farmers will always need ways to increase their productivity. Technology has brought new farm systems to bear. The automated thrasher can harvest a full field of corn faster than ever before. Wheat is also no problem. Personal computers can manage the farm finances and products costs at market. The modern farmer needs these technology tools or he will not make enough money to survive. Even with technology he will certainly look for government help in the form of a subsidy.

Figure 33. John Deere Farm Equipment[24]

The electronic farm has a generator for power and large trucks and tractors of all types. It has irrigation sprinkler equipment and harvesters. It requires the farmer to go to school in electronics and troubleshooting. Even schools here on the east coast once started as agricultural colleges such as Maryland University. So investment is required in electronics to run the modern farm. The larger farms may be near urban areas or markets in the megalopolis. The sure thing is that housing developments may crop up overnight depending on the weather and if a farmer can make a profit.

This has created a new ex-urbia in Maryland where once a housing development is built new services grow too along with the support systems they need for survival. Western Maryland is primarily farmland. Interstate 70 now has developments all along the highway system from Baltimore to Frederick, Maryland. People who work in Baltimore travel from Frederick so they can have affordable nice homes. They still need food so the farmer is needed as much as ever before. As growth continues more shopping centers are needed and the farmer has his newly suburbanized customers living near his farm. Food retailers know that this has happened and they are buying from closer sources to save

money.

24. Regionalism and Multi-State Operations

The BosWash megalopolis can be broken down into the smaller regions of three states or tri-state areas in the megalopolis. The northern most tri-state area is Massachusetts, Connecticut, and Rhode Island. Services in these three states may overlap and often they are considered New England. Boston, Hartford, and Providence are the major capital cities. Suburban structures now fill in the gaps between the CBD of each capital along the interstates. The rural areas can easily access the cities in a matter of hours.

The next tri-state area is New York, Connecticut, and New Jersey. The major metropolis is New York City which connects the three states by spanning from New Jersey to Connecticut. I-95 is again common to all three states. Furthermore New York is the gateway to Canada. It is the SMSA of New York City that reaches far north towards Hartford and south towards Philadelphia. Many people live in Connecticut and New Jersey and work in New York City. Sports teams are saturated in New York. The NFC East teams are from this area. The MLB National League East and American League East teams are from the BosWash megalopolis. Colleges may accept out of state students at in state tuition rates in these states. Many students from New York venture farther down the map to go to college in Maryland, DC, West Virginia, and Virginia. Again the interstates system reaches all those states within 4 -6 hours of driving. This also helps commercial traffic make the trips inside the BosWash megalopolis.

DelMarVa is the tri-state area of Delaware, Maryland, and Virginia. The Baltimore Washington metropolitan areas collide here. Most of Delaware on the Eastern Shore south is farmland but the north near Wilmington is very urban. I-95 runs through this part of the three states. Near borders people work in one state and live in another. People may shop in Delaware to avoid sales taxes. Companies locate in Delaware to avoid corporate taxes. Each state has it's own drawing card. In the summer each states tries to lure visitors from other neighboring states. Some may decide to buy property and live in both states.

The Pennsylvania, New Jersey, Delaware area is the fourth tri-state area. Shoppers may like to go to King of Prussia Mall near Philadelphia from the other states. Sports fans may go to teams in other nearby cities. Wilmington is at the center of this along with Philadelphia. The Jersey Turnpike and I-95 give plenty of interstate access to travelers. Commerce does well between these states. Dover AFB and McGuire AFB are located in this region. There are beaches in New Jersey and Delaware that Pennsylvania people can visit. And there are the Pocono mountains in Pennsylvania for New Jersey and Delaware people to visit.

The impact of these tri-state areas is economic. If one state has resources it can share with another then it does so in return for resources from the other state. In disaster and emergency times National Guard resources can also be shared. All of the states can survive on their own but it is to everyone's advantage to share resources. This makes regionalism a major force in the BosWash megalopolis. There is a larger force of the entire area but these sub states also have relationships that create new opportunities for people living there.

25. National Regulations

NIST (National Institutes of Standards and Technology) is the primary place we help define technology to the public. They produce standards in many technology areas and place those on their website at www.nist.gov. For computer systems they generate the FIPS PUBS which tell us how to build computer systems. They do not generate IEEE standards for communications. NIST also holds Biometric Conferences and they are the experts on computer security.

The FAA (Federal Aviation Administration) manages commercial aviation in the United States. FAA builds ATC systems and manages the air traffic. They regulate the commercial airlines industries for passenger planes and cargo planes. Every state aviation agency reports to FAA who controls the United States airways. State airspace and national airspace overlap so the agencies also overlap in authority and area of work. The local air networks feed into the national air networks. The publication that regulates FAA is the FAR (Federal Aviation Regulations).

The FCC (Federal Communications Commission) regulates communications including electronic and non-electronic communications in the United States. This includes radio frequencies and cable TV networks and any newer digital networks. They are key in understanding how we do business in the communications world. A frequency chart is available from NTIA (National Telecommunications Information Administration).

NHTSA (National Highway Traffic Safety Administration) regulates highway networks and commercial vehicles. FMCSA (Federal Motor Carrier Safety Administration) regulates truck commerce. These both have rules and regulations in the CFR system of regulations. States also have a role in regulating state highway traffic. Enforcement is done at the state and local police level for state law violations. In this regard, US Department of Transportation is a delegated agency to the states.

26. How Long will It Last – Judgement Day Cometh

Revelations Predictions include the fact that man will destroy the earth. Some say we are doing this by consuming all fossil fuels and earth resources faster and faster. The Bible tells us that Judgement Day is the day we must answer for our sins. Our systematic sins have been mainly creating pollution and destroying the environment during advanced systems building. Man's Nature is aggressive and disobedient. His warlike nature will end in Armageddon at a time in the future. The battle will be for good and evil. Jesus' Return will signify the second coming and peace for thousands of years. Hope will be returned to mankind.

Science Predictions say that man is definitely suing too many resources. The future holds times of vast resource deprivation. Carl Sagan has predicted we will run out of resources by 2030 and future generations will have great troubles because of current mismanagement of resources. Global Warming and Climate Change will be the straw that breaks the camel's back. Man will have to change his wasteful ways or entire cities will be consumed by rising water levels of the oceans. Green Fuels may help alter the future but we need to do more in this area. Vehicle emissions needs to be curbed. Electric cars need to become reality. Carbon based fuels need to be replaced. Our Destiny will be linked to the ability to respond to climate changes fast. Large scale engineering projects will not solve the coming problems.

Who Really Knows? The more Urban Systems we create the more we interact with the environment, the faster we head towards Judgement Day. If you take on the challenge of becoming an Urban Systems Engineer you must be sensitive to the problems we will face in the future. We must all become more environmentally aware of the outcomes of polluting the environment. I would say by 2050 we should know if we have done enough to change the future collision course with destruction of the environment. Major news items will hit the daily reports and the government will not be able to stop the scale of the disasters. They can not deal with major disasters now very well. Mother Nature seems to have the upper hand in hurricanes, tornados, melting icecaps, rising seawater, and global warming. Long before we began building any systems the earth had a balance to it and she compensated for climate changes for billions of years. Man needs to become much more sophisticated at handling our own earth climate Reacting to climate change may be the only response we have as we can not control climates yet. What would happen if we could make rain or stop hurricanes? This needs to be investigated further by the government.

## 27. Glossary and Acronyms

AC/DC – Alternating Current / Direct Current

ADA – Programming Language

AFB – Air Force Base

AMTRAK – Train Company

ARCVIEW – GIS viewer software developed by ESRI

ATC – Air Traffic Controller

AT&T – American Telegraph and Telephone

Baystat – Statistical Control System for Chesapeake Bay

BBC – British Broadcasting Company

BMC – Baltimore Metropolitan Council

BosWash – Boston to Washington

BWI – Baltimore Washington International

CBD – Central Business District

CFR – Code of Federal Regulations

Citistat – Statistical Control System for Baltimore City

COG – Council of Governments

CSX – Railroad company

CVISN – Commercial Vehicle Integrated System Network

CVO – Commercial Vehicle Operations

DARPA – Defense Advanced Research Projects Agency

DCDOT – District of Columbia Department of Transportation

DelMarVa – Delaware, Maryland, Virginia

DHS – Department of Homeland Security

DOD – Department of Defense

DOT – Department of Transportation

DSL – Digital Signal Line

ESRI – GIS company located in Towson, Maryland

Ex-Urbia – area outside of suburban areas from the city

FAA – Federal Aviation Administration

FAR – Federal Aviation Regulations

FBI – Federal Bureau of Investigation

FCC – Federal Communications Commission

FIOS – Fiber Optic System

FIPS PUBS – Federal Information Processing Standards Publications

FMCSA – Federal Motor Carrier Safety Administration

F-22 – Fighter Jet

GIS – Geographic Information System

GPS – Global Positioning System

GSA – General Services Administration

GW – George Washington

HDTV – High Definition Television

HDMI – High Definition Module Interface

ICCTV – Integrated Closed Circuit Television

IEEE – Institute of Electronics and Electrical Engineers

ISP – Internet Service Provider

ITS – Intelligent Transportation System

I-95 – Interstate 95

JFK – John Fitzgerald Kennedy

MASN – Middle Atlantic Sports Network

MCMIS –Motor Carrier Management Information System

Megalopolis – largest city of cities

MLB – Major League Baseball

NAVAIR – Naval Air Command

NAVAIDS – Navigational Aides

NAVIS – Navigational Information System

NextGen – Next Generation Air Traffic Control System

NFC – National Football Conference

NFL - National Football League

NG – National Guard

NHTSA –National Highway Transportation Safety Administration

NTIA – National Telecommunications Information Administration

NIST – National Institutes of Standards and Technology

NYC – New York City

PC – Personal Computer

PV – Photo Voltaic

RFID – Radio Frequency Identifier

Rural – agricultural farmlands

SAFER – Safety and Fitness Electronic Records System

SCADA – Supervisory Control And Data Acquisition

SMSA – Standard Metropolitan Statistical Area

SOC – State Operations Center

StateStat – Statistical Control System for Maryland

STC – Software Technology Center

Suburban – area just outside of Central Business District with residents

TSA – Transportation Security Agency

T1 – Telecommunications channel

UAV – Unmanned Aerial Vehicle

VOIP – Voice Over Internet Protocol

WGN – Chicago cable TV station

WIM – Weigh In Motion

## 28. Bibliography

Books

Daganzo, Carlos, (1997), <u>Fundamentals of Transportation and Traffic Operations</u>, Pergamon, Elsevier Science Inc., New York.

Fink, D, and Beaty, W, (1993), <u>Standard Handbook for Electrical Engineers</u>, 13th Ed., McGraw-Hill, New York.

Gillmer, T & Johnson, B, (1982), <u>Introduction to Naval Architecture</u>, Naval Institute Press, Annapolis, Maryland.

Gist, Noel P. & Fava, Sylvia F., <u>Urban Society</u>, 6th Ed., Thomas Cromwell Company, 1974, New York.

Homburger W., Hall J., Reilly W., Sullivan E., (2001), <u>Fundamentals of Traffic Engineering,</u> 15th Ed., Institute of Transportation Studies, Univ. California, Berkley.

McKeever, Ross, (1968), <u>The Community Builders Handbook</u>, Urban Land Institute, Washington DC.

Menzefricke, Ulrich., (1995), <u>Statistics for Managers</u>, Duxbury Press, Wadsworth Publishing Company, Belmont, California.

Meyer, Michael and Miller, Eric, (1984), <u>Urban Transportation Planning: A Decision-Oriented Approach</u>, McGraw-Hill, New York.

Stallings, William, (2005), <u>Business Data Communications</u>, 5$^{th}$ Ed., Pearson

Prentice Hall, Upper Saddle River, New Jersey.

Tanenbaum, Andrew, (1981), <u>Computer Networks</u>, Prentice Hall, New Jersey.

Taneja, Nawal, (1976), <u>The Commercial Airline Industry</u>, Lexington Books,

Massachusetts.

US Congress – 91$^{st}$ (1968), <u>Building The American City; Report of the National

Commission on Urban Problems to the Congress and to the President of

the United States</u>, US Government Printing Office, Washington DC.

## Websites

www.google.com
www.wikipedia.com
www.co.ho.md.us
www.mdot.state.md.us
www.nyc.gov/dot
www.cite.edu
www.umd.edu
www.builditsolar.com
www.ecoplan.org/kyoto/challenge/traffic.htm
www.googleearth.com
www.mapquest.com
www.faa.gov
nab.usace.army.mil
www.onebigcampus.com
www.noaa.gov
www.johndeere.com
www.perronerobotics.com

29. Biography


Donald Joseph Gray Chiarella lives in Elkridge-Hanover in Howard County, Maryland with is wife Mimi, a master teacher, and 4 great children. Born in Kilmarnock, Scotland in 1956 to a US Air Force family, eldest son of Donald Sr. and Margaret of Orlando, Florida. Don is now a Homeland Security Coordinator in the Motor Carrier Division of Maryland SHA and has been with the State of Maryland for 12 years. He was previously the MIS Section Chief (G-19) for Maryland State Highway Administration in the Traffic Safety Analysis Division from 1997-2006. He won Division Employee of the Year in 2005. He is in the 1997-2007 editions of Who's Who in America. He holds an independent study online Ph.D. from Kennedy-Western University in MIS (2001 - unaccredited), an M.S. degree in Technology of Management from American University (1988) (Dean's List), and a B.A. degree in Urban Planning / IFSM from University of Maryland (1979) with the first URBS degree in the IFSM specialty. He was Vice President of DPMA at American University. He has post graduate study in Transportation Engineering and Systems Engineering from University of Maryland Clark School of Engineering. He also holds a second bachelors degree from Ashford University online in Organizational Management (2009) with distinction and a 3.85 GPA. He received a Master Certificate from George Washington University (1996) in Public Contracts Management. He is also certified by the ICCP and DAMA as a Certified Data Management Professional (CDMP). He is a Certified Information Security Manager by ISACA (CISM). He has attended the Naval Post Graduate School Homeland Defense

Courses, National Defense University Information Management, Department of Defense Computer Institute Project Management, American Military University Anti-Terrorism, and US Air Force Academy Dooley Year.  Previously he worked as a mainframe computer specialist (GS-12) 10 years for the US Navy Medical Data Services Center at Bethesda and 7 years (GM-13) for GSA Central Office in Washington DC before retiring from Federal Civil Service.  He served as Treasurer of the Professional Managers Association in Washington. He led the GSA FAR/FIRMR CDROM Project to 1994 CDROM of the Year Award in Washington DC.  He helped manage installation and build the GSA Token Ring Netware LAN.  He wrote the FIRMR policy that managed all outdated computers and CASE Tools in federal government.  He led consultant systems analyst teams of Natural/ADABAS/CLIPPER database experts from 1987-1990 on five different agency projects in Washington DC ranging from Army Intelligence/JPL networking, Naval Ordnance Civilian Personnel, Pentagon Foreign Counter Intelligence Budget Office, OPM Retirements Division, and FBI Headquarters Indexing Name System. He won two bonus increases during this timeframe.  He has held a Secret Clearance.  He is a 1974 Nixon presidential and 1975 Maryland state scholarship winner and has won a college championship in baseball (catcher) in the PIC NAIA Conference in 1976 at St. Mary's College of Maryland.  He is ASEP coaching certified and served as a little league commissioner for the Upper Marlboro Boys and Girls Club in basketball in 1987. He won an award for staff mentoring from GSA in 1994.  He is a life member of the US Naval Institute, AFCEA, American University Alumni, and USAF Academy Association of Graduates. He is also a member of NARFE, AARP, and AFT/MPEC.  He was Central high school athlete of the year in 1974 (baseball, basketball, football) and All County in Baseball. He won PG County Rotary Club student of the month award and Kiwanis student of the month awards. He was in the National Honor Society. Don has personally built over 50 computer mainframe and micro software systems, some networks, webpages, CDROMS, and management software and written many papers and documents for government systems management and planning and the private sector. He has studied 4 foreign languages and more than 10 computer languages, 7 operating systems, and 5 databases. He was a Democratic Chief Voting Judge in Howard County securing electronic computer voting machines. He is on staff at Aspen University in Denver, Colorado a US Dept of Education DETC certified IT school teaching Data Communications. He also teaches Computer Security at George Washington University Arlington Professional Studies campus. He has taught over 25 college and high school courses and written 15 books over the years some of which are available at www.lulu.com/donchiarella. His favorite charities are the Maryland Masons, Compassion International, United Methodist Church, and Fellowship of Christian Athletes. He enjoys travel, sports, golf, motor cycling, teaching, reading, bible study fellowship, UM men's group, gardening, and family time.

This page left intentionally blank.

www.ingramcontent.com/pod-product-compliance
Lightning Source LLC
Chambersburg PA
CBHW051415200326
41520CB00023B/7240